Emergency Response and Recovery Planning for Water Systems: A Kit of Tools

Awwa Research Foundation

About the Awwa Research Foundation

The Awwa Research Foundation (AwwaRF) is a member-supported, international, nonprofit organization that sponsors research to enable water utilities, public health agencies, and other professionals to provide safe and affordable drinking water to consumers.

The Foundation's mission is to advance the science of water to improve the quality of life. To achieve this mission, the Foundation sponsors studies on all aspects of drinking water, including supply and resources, treatment, monitoring and analysis, distribution, management, and health effects. Funding for research is provided primarily by subscription payments from approximately 1,000 utilities, consulting firms, and manufacturers in North America and abroad. Additional funding comes from collaborative partnerships with other national and international organizations, allowing for resources to be leveraged, expertise to be shared, and broad-based knowledge to be developed and disseminated. Government funding serves as a third source of research dollars.

From its headquarters in Denver, Colorado, the Foundation's staff directs and supports the efforts of more than 800 volunteers who serve on the board of trustees and various committees. These volunteers represent many facets of the water industry, and contribute their expertise to select and monitor research studies that benefit the entire drinking water community.

The results of research are disseminated through a number of channels, including reports, the Web site, conferences, and periodicals.

For subscribers, the Foundation serves as a cooperative program in which water suppliers unite to pool their resources. By applying Foundation research findings, these water suppliers can save substantial costs and stay on the leading edge of drinking water science and technology. Since its inception, AwwaRF has supplied the water community with more than $300 million in applied research.

More information about the Foundation and how to become a subscriber is available on the Web at **www.awwarf.org.**

Emergency Response and Recovery Planning for Water Systems: A Kit of Tools

Prepared by:
Charles Herrick, **Joanna Pratt**, **Robert Raucher**, and **Nicole Kalas-Adams**
Stratus Consulting Inc.
P.O. Box 4059, Boulder, CO 80306-4059

Joseph Cotruvo, J. Cotruvo Associates

Kristin Darr-Bornstein, KDA Creative

Timothy Gablehouse, Gablehouse Calkins & Granberg, LLC

John Mencer, Infrastructure Protection Group

Richard Moser, RHMoser Consulting

Dale Newkirk, Damon S. Williams Associates

Mary Schnack, Mary Schnack Media Services, Inc.

and

Kennet Westby, Coalfire Systems, Inc.

Sponsored by:
Awwa Research Foundation
6666 West Quincy Avenue, Denver, CO 80235-3098

Published by:

Awwa Research Foundation

DISCLAIMER

This study was funded by the Awwa Research Foundation (AwwaRF). AwwaRF assumes no responsibility for the content of the research study reported in this publication or for the opinions or statements of fact expressed in the report. The mention of trade names for commercial products does not represent or imply the approval or endorsement of AwwaRF. This report is presented solely for informational purposes.

CONTENTS

EXHIBITS

TABLES

FIGURES

FOREWORD

The Awwa Research Foundation is a nonprofit corporation that is dedicated to the implementation of a research effort to help utilities respond to regulatory requirements and traditional high-priority concerns of the industry. The research agenda is developed through a process of consultation with subscribers and drinking water professionals. Under the umbrella of a Strategic Research Plan, the Research Advisory Council prioritizes the suggested projects based upon current and future needs, applicability, and past work; the recommendations are forwarded to the Board of Trustees for final selection. The foundation also sponsors research projects through the unsolicited proposal process; the Collaborative Research, Research Applications, and Tailored Collaboration programs; and various joint research efforts with organizations such as the U.S. Environmental Protection Agency, the U.S. Bureau of Reclamation, and the Association of California Water Agencies.

This publication is a result of one of these sponsored studies, and it is hoped that its findings will be applied in communities throughout the world. The following report serves not only as a means of communicating the results of the water industry's centralized research program but also as a tool to enlist the further support of the nonmember utilities and individuals.

Projects are managed closely from their inception to the final report by the foundation's staff and large cadre of volunteers who willingly contribute their time and expertise. The foundation serves a planning and management function, and awards contracts to other institutions such as water utilities, universities, and engineering firms. The funding for this research effort comes primarily from the Subscription Program, through which water utilities subscribe to the research program and make an annual payment proportionate to the volume of water they deliver and consultants and manufacturers subscribe based on their annual billings. The program offers a cost-effective and fair method for funding research in the public interest.

A broad spectrum of water supply issues is addressed by the foundation's research agenda: resources, treatment and operations, distribution and storage, water quality and analysis, toxicology, economics, and management. The ultimate purpose of the coordinated effort is to assist water suppliers to provide the highest possible quality of water economically and reliably. The true benefits are realized when the results are implemented at the utility level. The foundation's trustees are pleased to offer this publication as a contribution toward that end.

Walter J. Bishop
Chair, Board of Trustees
Awwa Research Foundation

Robert C. Renner, P.E.
Executive Director
Awwa Research Foundation

ACKNOWLEDGMENTS

This report was funded by the Awwa Research Foundation. The project was managed and administered by Ms. India Williams. The researchers would like to extend their great appreciation to AwwaRF, and especially to India for her management, patience, and unflagging support and encouragement at every stage of the project. We also wish to thank the project advisory committee (PAC), which provided useful comments, guidance, and critiques at various stages of this project:

Albert Perez, Public Utilities Department, City of North Miami Beach, Fla.
Kenneth Stone, U.S. Environmental Protection Agency, National Homeland Security Research Center
John Spatz, Bureau of Water Supply, Chicago Department of Water

The researchers would like to thank the participating utilities for their input and feedback on all aspects of the research, analysis, and reporting of this work. Specifically, we would like to thank:

City of Portland Bureau of Water Works
District of Columbia Water and Sewer Authority
Department of the Army Washington Aqueduct
Sonoma County Water Agency
San Francisco Public Utilities Commission Water Quality Bureau
American Water
Fairfax County Water Authority
South Central Connecticut Regional Water Authority
Hidden Valley Lake Community Services District
Alameda County Water District
Washington Suburban Sanitary Commission

We would also like to thank the research team, individuals who attended the project workshop in San Francisco, and the utilities that participated in our survey process.

Last, but not least, the researchers wish to thank the many individuals who helped execute the technical tasks and produce this report, especially Diane Callow, Laura Cross, and Erin Miles of Stratus Consulting, René Howard of WordProse, Inc., Christina Thomas, and Christine Teter.

Thank you all very much.

Charles Herrick, PhD (Principal Investigator)
Stratus Consulting Inc.
Boulder, Colo. (May 20, 2005)

EXECUTIVE SUMMARY

One consequence of the events of Sept. 11, 2001, is a heightened concern among U.S. citizens about the security of their critical drinking water infrastructure. With 168,000 public water systems in the United States, the potential for terrorists to damage the nation's water infrastructure remains considerable.

Government agencies and water utilities already have significant mechanisms in place for managing a wide range of emergencies and disasters, including earthquakes, hurricanes, disease epidemics, and accidental releases of toxic substances into the environment. The types of events now being contemplated, however, are different. When an event occurs that is so rare and damaging that it outstrips preparations, plans, and social and material resources, severe problems arise. These new threats can include terrorism, sabotage, and cyberattacks. Some argue that the tenets of conventional emergency planning and management are no longer sufficient to protect water utilities from the range and nature of threats that have come to be conceivable in the United States.

RESEARCH OBJECTIVES

The broad objectives of Awwa Research Foundation Project Number 2929 are to improve the state of the practice of disaster response, recovery, and business continuity planning for water utilities; and to provide usable planning tools that help water utilities update existing capabilities and address new contingencies such as domestic terrorism. Specifically, the project has four basic research objectives:

- Identify and review existing guidance and other resources on emergency response planning
- Characterize the current state of the practice for water utility emergency response planning
- Conduct expert and stakeholder forums to present findings and obtain feedback
- Identify gaps in and best practices for water utility emergency plans and planning programs

The ultimate objective of the project is to integrate and synthesize research findings to develop a user-friendly kit of tools for utilities to use as they develop and/or update their emergency plans and overall security programs.

APPROACH

For this project, the Stratus Consulting research team used an ensemble research methodology to integrate findings from a series of topically focused literature reviews, an in-depth survey of utilities, the team's content analysis and critical review of utility plans and water sector guidance materials, and a stakeholder and expert workshop. As the first step, the research team collected and critically reviewed existing resources on emergency response/recovery and business continuity planning. Specifically, this step included:

- Review of existing emergency response guidance materials developed for the water sector
- Inventory of existing software tools for emergency planning
- Legal analysis of the Public Health Security and Bioterrorism Preparedness and Response Act of 2002 (Public Law 107-188, commonly called the Bioterrorism Act), and implications for the water sector
- Review of the literature in the areas of disaster response, recovery, and business continuity planning

Next, the research team characterized the current state of the practice of emergency and business continuity planning in the water sector. This step included

- Survey of water utilities to ascertain emergency response planning practices
- Review of existing water utility emergency response plans
- Literature review focused on business continuity planning, within and outside the water sector
- Literature review focused on emergency communications
- Review of existing utility mutual assistance agreements
- Roundtable discussion with the Greater Washington Council of Governments to discuss informal approaches to mutual assistance

As a third step, the research team convened a stakeholder and expert workshop to brief utilities on interim findings, elicit utility input on research focus areas and feedback on proposed project outputs, and obtain best practices information. In particular, input was sought on the following topics:

- Development of disaster and emergency scenarios
- Stakeholder identification and characterization
- Business continuity planning
- Training and drills and plan acculturation
- Emergency communications plans
- Mutual assistance agreements
- Software tools to support emergency planning

As a fourth step, the research team assessed the information gathered from these activities to identify best practices and gaps in emergency response planning. The final step was to develop a framework to help utilities systematically and formally review their own emergency response plans to determine whether they suffer from the types of gaps identified through project research. In addition, the framework will assist utilities in determining whether their plans are responsive to the Bioterrorism Act's requirements.

CONCLUSIONS

The research conducted for this project yielded a number of important findings, and nearly all have practical implications for how water utilities should develop and update emergency response plans. In addition, and equally important, the study's findings should be used to inform utility efforts to "enliven" their paper plans with a corporate culture conducive to a "mindful" approach to emergency preparedness and response. A sampling of key project findings follows:

- The Bioterrorism Act treats vulnerability assessment and emergency planning in the water sector as ongoing priorities, not as a one-time "paper" exercise.
- In comparison to natural disaster phenomena, the possibility of domestic terrorism introduces a new, relatively unpredictable element to water security planning in the United States. This fundamental lack of predictability underscores a need to augment plans and pre-defined response protocols with training to develop and enhance operator mindfulness and real-time emergency problem-solving skills.
- Existing water system emergency response plans appear uneven in their reliance on formal emergency scenarios. Although some include scenario-specific action protocols, most do not formally consider particular scenarios in determining resource needs, action priorities, or other planning elements. In addition to adding an important element of specificity to a utility's emergency response and recovery plan, the scenario development process is itself a valuable exercise in helping utility staff to recognize and swiftly respond to emergency situations.
- Many water utilities emphasize the importance of training, but response plans suggest that systems vary widely in their training schedules, and in the degree to which training exercises involve utilization of the plan itself.
- Although a few utilities are beginning to grapple with the advanced concept of business continuity planning, most have not taken steps to prioritize the needs of their customer base and develop corresponding thresholds of service disruption.
- Many water system emergency plans appear not to be based on formal consideration and analysis of how different emergency scenarios will affect different stakeholder groups. Many emergency plans merely list "typical" or "likely" stakeholders, but fail to address how the same event may result in radically different impacts on different stakeholder groups.
- Many utility emergency response plans are voluminous, complex, and appear difficult to implement; more than half the utilities surveyed under this project characterize their plans as cumbersome. The "usability" of emergency plans under actual (stressful, time-limited) emergency conditions is a matter of concern.

Most of these conclusions represent gaps or apparent insufficiencies in current water utility plans and programs. An approach for addressing these gaps is described in the next section.

RECOMMENDATIONS

The kit of tools produced under this project is designed to help utilities conduct critical self-assessments of their existing plans and associated security programs. Central to the kit of tools is a heuristic framework based on a sequential characterization of actions likely to unfold during an emergency situation. The framework consists of 12 "action areas," each of which incorporates a series of questions designed to help utilities periodically evaluate the degree to which their plans and emergency programs are adequate to address known and emerging threats and vulnerabilities. The kit of tools includes a total of 73 self-assessment questions.

The kit of tools is contained within the CD-ROM enclosed. It is configured so that utilities can identify areas of potential concern and then quickly and easily obtain guidance, best practice information, and other background materials to help them address identified issues. Resources contained on the CD-ROM were either developed specifically for this project or excerpted from quality-assured professional sources.

This report also addresses a series of special focus areas important for water sector emergency planning. Dealing with topics such as stakeholder analysis, emergency communication, emergency plan usability, and business continuity planning, these discussions direct users to relevant aspects of the CD-ROM and present additional information and guidance.

CHAPTER 1
PURPOSE AND APPROACH

PURPOSE

This report contains a written document and an accompanying CD-ROM that together comprise a "kit of tools" designed to assist community water systems with preparing and maintaining their emergency response and recovery plans and ongoing emergency response and preparedness programs. It is intended to help water systems organize their planning efforts, and to provide a one-stop reference for the types of information that an emergency response plan should include. This document and supporting materials can be used by all water systems, regardless of size.

The tools provided in this report are intended to assist utilities in planning for all types of emergency scenarios, including natural disasters and human-caused emergencies. Emphasis is placed, however, on planning for risks related to terrorism, because domestic terrorism constitutes a new and growing concern for water utilities.

REPORT ORGANIZATION

The document is composed of five major parts. Chapter 1 describes the study purpose and organization, and presents the five-step methodology that governed our research and analysis. Chapter 2 contains background information on terrorism and emergency response planning, and presents some of our key research findings. More specifically, the chapter provides background on terrorism in the United States (focusing on potential threats to public water systems and other water utilities). As a guide to the remainder of the document, chapter 2 presents a lexicon of emergency planning and disaster response terminology that has been drawn from seminal works and key guidance. The chapter also contains a summary of the Public Health Security and Bioterrorism Preparedness and Response Act of 2002 (Public Law 107-188, subsequently referred to as the Bioterrorism Act), outlining specific actions that utilities are required to undertake in an effort to enhance security and address potential terrorist threats. A summary of research undertaken in support of this project is presented, including an in-depth survey of water utilities to assess the current status of emergency response planning. Perhaps most importantly, a research findings summary articulates a series of *gaps* in the water sector's approach to emergency response and recovery planning.

Chapter 3 is configured as a kit of tools, organized to help water systems conduct self-assessments to determine the degree to which their emergency response plans and programs are (1) consistent with the stipulates and reasonable implications of the Bioterrorism Act, and (2) generally consistent with best practices in emergency planning. Chapter 3 is organized around a heuristic framework and a series of 73 "self-assessment questions." The self-assessment questions correspond to items in a list of references, enabling utility staff and partners to easily access appropriate "how to" guidance, templates, and illustrative best practice information. The kit of tools is intended to allow water system personnel to quickly locate guidance and best practices that address gaps or weaknesses in their current planning approach. It is intended to be used in conjunction with a CD-ROM of resources, described in the section that follows.

Chapter 4 provides information on seven special focus areas that are critical to water utility emergency planning and response. The information presented in this chapter is based on targeted literature reviews conducted by the research team and input elicited through expert workshops and roundtable discussions. The special focus areas include (1) disaster scenarios; (2) stakeholder identification and role definition; (3) business continuity planning; (4) emergency response training and drills; (5) communications guidance, strategies, and protocols; (6) mutual assistance agreements and approaches; and (7) software packages that support and automate emergency preparedness and response.

Chapter 5 presents recommendations on configuring and structuring an emergency response plan. It includes suggestions for ensuring that plans are designed so they can be used easily and efficiently under high-stress emergency conditions.

CD-ROM ORGANIZATION

The CD-ROM included with this report was developed for use in conjunction with the kit of tools presented in chapter 3. It contains best practice materials excerpted from quality-assured, professionally recognized sources of emergency response planning guidance. It also contains guidance materials developed or adapted by the Stratus Consulting research team to address gaps or insufficiencies identified through background research. These resources are numerically indexed to the framework and self-assessment questions introduced in the kit of tools.

RESEARCH APPROACH

This project was based on the five-step research approach illustrated in Figure 1.1. As the first step, the research team collected and reviewed existing resources on emergency response/recovery and business continuity planning. This step included:

- Compilation and review of existing emergency response, recovery, and business continuity planning guidance materials developed by and for the water sector
- Inventory of existing software tools and Web applications for emergency planning and response, and for business continuity
- Legal analysis of the requirements of the Bioterrorism Act and implications for the water sector
- Literature review of academic and professional studies in the areas of disaster response, recovery, and business continuity planning, both within and outside the water sector

Second, Stratus Consulting and its research partners characterized the current state of the practice of emergency and business continuity planning in the water sector. This aspect of our research was based on an in-depth survey of selected utility officials, a series of targeted literature reviews, and review of individual utility emergency response plans. Note that in all cases, the research team was granted review of utility plans, documents, and other materials only on the condition of strict confidentiality and source anonymity. For this reason, this study does not name or cite any specific utilities or organizations.

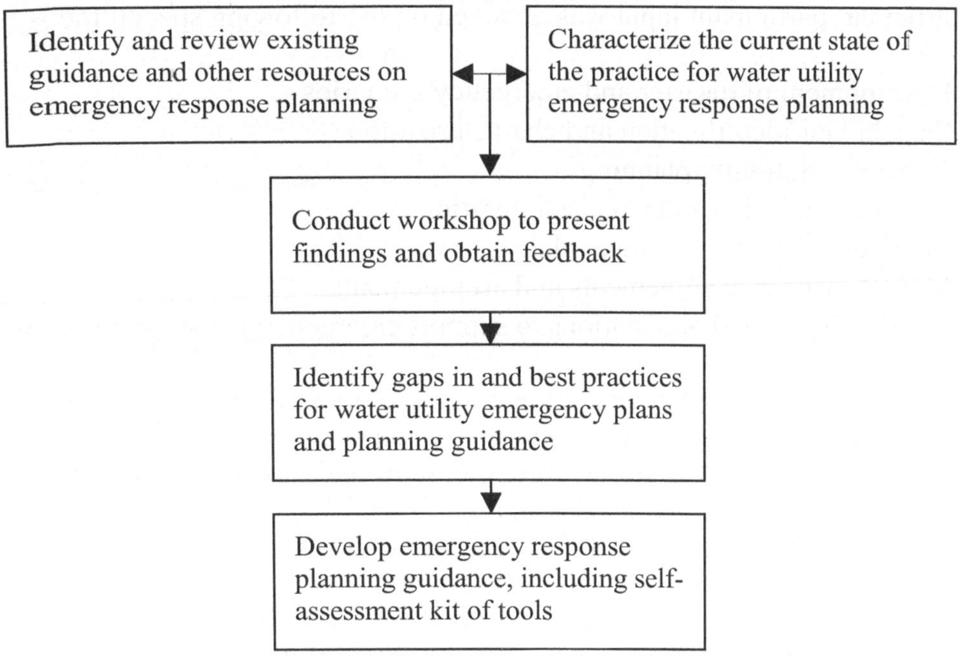

Figure 1.1 Emergency response, recovery, and business continuity planning research approach

Specifically, the Stratus Consulting research team conducted the following reviews:

- Survey of water utilities to ascertain their emergency response and recovery and business continuity planning practices, approaches, needs, strengths, and limitations
- Literature review of existing water utility emergency response plans
- Literature review of the status of business continuity planning, both inside and outside the water sector
- Literature review of existing emergency and disaster response communications guidance and plans conducted by the water sector and other industries
- Critical review of existing water utility and municipal mutual assistance agreements
- Roundtable discussion on mutual assistance agreements held with water utility members of the Greater Washington Council of Governments (COG) to address their approach to mutual assistance

The third step of our research approach was to conduct a workshop on emergency response, recovery, and business continuity planning for water utilities. The key objectives of the workshop, held July 8–9, 2004 in San Francisco, were to (1) brief utility experts on our interim findings and emerging perspectives (from steps 1 and 2); (2) elicit utility input on the practicality, applicability, and relevance of our project findings and proposed outputs; and (3) obtain additional information from the participants on best practices for emergency response/recovery and business continuity planning.

In particular, participant input was obtained on the following special focus areas:

- Development of disaster and emergency scenarios
- Stakeholder identification and characterization
- Business continuity planning
- Training and drills and plan acculturation
- Emergency communications plans
- Mutual assistance agreements and arrangements
- Software tools and applications to support emergency response planning

Workshop participants included a wide range of experts and practitioners with extensive experience in water utility disaster response and recovery planning. About half of the 30 participants were representatives from large, medium, and small/rural drinking water utilities. Also present were several individuals from the regulatory community, including the U.S. Environmental Protection Agency (USEPA) and the California Department of Health Services, along with members of the Stratus Consulting research team.

The fourth step of the research process involved assessing the information obtained from steps 1 through 3 to identify (1) best practices in the arena of water system security planning and (2) gaps in current approaches to emergency response/recovery and business continuity planning and/or planning guidance. Key conclusions are presented in the remainder of this report.

Last, the Stratus Consulting research team developed a framework and associated guidance materials to help utilities quickly—although formally and systematically—review their own emergency response/recovery and business continuity plans to assess (1) whether they suffer from the types of gaps identified through this project, and (2) whether their plans seem consistent with industry practice and guidelines. The final product includes:

- A kit of tools for water utilities to use in assessing their existing emergency response/recovery plans or developing their initial plan (see chapter 3 and the CD-ROM included with this report)
- Focused guidance for special areas of emergency response/recovery planning (see chapter 4)
- Guidance on how to structure and organize an emergency response plan (see chapter 5)

CHAPTER 2
BACKGROUND AND SUMMARY OF FINDINGS

This chapter presents background information on the Bioterrorism Act and its implications for water utility emergency response planning. It also includes an assessment of the current emergency response planning state of the practice among water utilities, based on a utility survey conducted for this project. Based on information from this survey and other research activities, we also present a list of gaps in the water sector's approach to emergency response and recovery planning. As a reference guide for this chapter and the remainder of the report, Exhibit 2.1 provides a lexicon of key emergency response planning terminology.

Exhibit 2.1
Emergency planning terminology

Emergency. An exceptional event that exceeds the capacity of normal resources and organization to cope with it (Alexander 2002). A water service emergency is a forced water shortage due to inadequate supply, deterioration of quality, or damage and partial isolation (Mutual Assistance Agreement 1982).

Incident. Any situation with the potential to disrupt essential services, or jeopardize public health and safety (Metropolitan Washington COG 2002).

Mitigation. Actions taken to reduce the impact of future disasters (Alexander 2002).

Preparedness. Actions taken to reduce the impact of disasters when they are forecast or imminent (Alexander 2002).

Response. Emergency actions taken during both the impact of a disaster and the short-term aftermath (Alexander 2002).

Recovery. The process of repairing damage, restoring services, and reconstructing facilities after disaster has struck (Alexander 2002).

Business continuity. The documentation of strategies, procedures, and resources used by an organization to respond to, recover from, and resume operations in the event of a substantial disruptive incident (Moore 2000).

THE BIOTERRORISM ACT AND WATER UTILITY RESPONSIBILITIES

In response to the tragic events of September 11, 2001, Congress determined that there was a need to mobilize resources of the United States to enhance security and improve our ability to respond to the threat of domestic terrorism. As a key element of this congressional response, the Bioterrorism Act covers a great deal of new ground and creates new programs and authorities to deal with issues ranging from emergency response to food safety.

The basic requirements of the Bioterrorism Act of concern to community water systems are found in its amendments to the Safe Drinking Water Act (SDWA). A new section 1433 is added to the SDWA and is now found as 42 USC 300i-2. This section contains two aspects of particular importance (summarized in Exhibit 2.2):

1. Community water systems serving more than 3,300 persons must conduct a vulnerability assessment. Of note is the fact that the statute focuses this vulnerability assessment on terrorist or intentional acts that are intended to disrupt the ability of the system to deliver safe and reliable supplies of water. The statute provides that the assessment must consider pipes and conveyances, physical barriers, the entire process of water collection through delivery to customers, automated systems, use and storage of chemicals, and maintenance. The deadlines for completing this work have passed.

2. Once the vulnerability assessment is completed, the Bioterrorism Act gave community water systems serving more than 3,300 people six months to certify to the USEPA that they have completed an emergency response plan addressing vulnerabilities. Where possible, these plans were to have been completed in coordination with Local Emergency Planning Committees (LEPCs). (For background on LEPCs see http://www.epa.gov/ceppo/lepclist.htm.) The emergency response plans must identify plans, actions, procedures, and equipment that can be used to stop or significantly reduce the impact of a terrorist or intentional act, and to respond to such an act should it occur.

Exhibit 2.2
Bioterrorism Act emergency planning requirements

If a water utility serves a population greater than 3,300, it shall prepare or revise an emergency response plan that

- Incorporates results of its vulnerability assessment

And includes, but is not limited to:

- Plans, procedures, and identification of equipment that can be used in the event of a terrorist attack
- Plans, procedures, and identification of equipment that can be used to lessen the impact of a terrorist attack

As possible, water systems shall coordinate with Local Emergency Planning Committees (as established under the Emergency Planning & Community Right to Know Act [EPCRA]).

The USEPA has developed various guidance materials and tools to assist utilities in their efforts to comply with the Bioterrorism Act. Utilities that have yet to develop or update their emergency response plan are urged to obtain and adopt USEPA guidance. Exhibit 2.3 summarizes guidance, reference materials, and other resources that the USEPA makes available to community water systems.

The Bioterrorism Act does not specifically state whether a community water system is required to implement all the procedures, buy all the equipment, and/or make all of the changes that might be suggested through the vulnerability assessment and emergency response planning process. Although the act itself is silent on the extent to which systems are expected to implement mitigation measures and related emergency procedures, some guidance on questions such as these can be gleaned through review of the SDWA itself. Especially pertinent in this context is SDWA Section 1431, 42 USC 300i, as amended by the Bioterrorism Act, which gives the USEPA the authority to issue orders to any water supplier that fails to take "adequate steps to prevent terrorist or intentional acts if an imminent and substantial endangerment to public health is present." Even though it is difficult to determine exactly how this provision might be applied, it is clear that community water suppliers have a duty to take those steps necessary to eliminate imminent and substantial endangerment to public health from terrorism or intentional acts. Also relevant is new section 1435 of the Bioterrorism Act, found at 42 USC 300i-4, which requires the USEPA to study the techniques terrorists or others might actually use to cause harm to U.S. water systems. The agency must then evaluate and recommend protective steps to prevent this harm, and recommend procedures by which alternative water supplies could be made available.

The USEPA's activities anticipated under Sections 1431 and 1435 imply that Congress intends for vulnerability assessments and emergency response plans to be updated with some degree of regularity. In other words, Congress has recognized that the terrorist threat to the United States is fluid and likely to evolve over time. Emergency planning, as stipulated under the Bioterrorism Act and SDWA, thus denotes an ongoing process, not a single-point-in-time event or product. Based on this recognition, it seems prudent for water systems to periodically reassess and update their emergency response plans to keep up with the evolving threat of domestic terrorism.

Exhibit 2.3
USEPA water security guidance*

- Emergency Response Plan Guidance for Small and Medium Community Water Systems to Comply with the Public Health Security and Bioterrorism Preparedness and Response Act of 2002 (U.S. EPA 2004)
- Instructions to Assist Community Water Systems in Complying with the Public Health Security and Bioterrorism Preparedness and Response Act of 2002 (U.S. EPA 2003a)
- Guidance for Water Utility Response, Recovery & Remediation Actions for Man-Made and/or Technological Emergencies (U.S. EPA 2002) (http://www.epa.gov/safewater/watersecurity/pubs/er-guidance.pdf)
- Large Water System Emergency Response Plan Outline: Guidance to Assist Community Water Systems in Complying with the Public Health Security and Bioterrorism Preparedness and Response Act of 2002 (U.S. EPA 2003b)
- Public Notification Handbook (U.S. EPA 2000) (http://www.epa.gov/safewater/pws/pn/handbook.pdf)
- Response Protocol Toolbox: Planning for and Responding to Drinking Water Contamination Threats and Incidents (Six "Modules," Interim Final Version, December, U.S. EPA 2003c-i)

  Module 1: Water Utilities Planning Guide
  Module 2: Contamination Threat Management Guide
  Module 3: Site Characterization and Sampling Guide
  Module 4: Analytical Guide
  Module 5: Public Health Response Guide
  Module 6: Remediation and Recovery Guide

*Unless otherwise noted, these documents can be found at http://cfpub.epa.gov/safewater/watersecurity/.

THE NATURE OF THE TERRORIST THREAT: EMERGENCY RESPONSE PLANNING IMPLICATIONS

Utilities must prepare for a range of both natural and human-caused disasters (see Exhibit 2.4). As emergency management expert David Alexander makes clear, natural disasters tend to be somewhat repetitive and to concentrate in particular areas. For example, seismic and volcanic belts, hurricane-generating areas, tornado "alleys," flood plains, and areas susceptible to wildfire are well documented, and the statistical frequency of many natural hazard events is fairly well established. Similarly, some types of catastrophic technological failures are also subject to statistical regularity. Terrorist attacks, however, especially those carried out on specific U.S. domestic resources, do not occur with any type of statistical regularity. This makes them exceedingly difficult to predict, even in the abstract (Alexander 2002).

Many experts argue that the tenets of classical crisis management no longer apply in a post 9–11 world. Many organizations, including water utilities, develop and refine plans to improve the way they handle emergencies of the kind they have faced in the past, but they may face a bigger threat from calamities they have never encountered, or even worse, from threats they cannot even imagine. As Mitroff and Alpaslan (2003) write, we are essentially "fighting new wars with old strategies." Terrorism expert Richard Little argues that we "find ourselves in a time where former contexts of threat, vulnerability, and target have all changed and continue to do so. Threats are radically unpredictable and the full range of targets probably unknowable. We can never anticipate all possible threats and even if we could, we lack the resources to deploy technologies to address them" (Little 2004). Security in such a situation must be flexible, capable of recognizing and responding to new threats as they emerge. Utilities are investing millions of dollars in various protective technologies and risk mitigation measures, ranging from fencing and locks to high-technology networks for monitoring distribution systems. But such technologies will only make water systems safer if they are supported by organizations and people prepared to manage the response to and recovery from an attack.

Exhibit 2.4
Categories of hazards

Natural disasters

- Earthquake
- Flooding (river, flash, coastal, dam break)
- Wind (hurricanes, tornados)
- Drought
- Other severe weather (cold, heat, snow, ice)
- Fire (forest, brush)
- Volcano
- Mudflow and landslide

Human-caused disasters

- Hazardous material releases
- Breaks, system failures
- Major accidents (construction, transportation)
- Structure fires
- Terrorism
- Vandalism
- War and civil unrest

Source: Adapted from Grigg 2002.

Utilities should therefore converge on emergency planning from two different perspectives: a macro-design mode and a micro-operational level. In the macro mode, formal, deductive reasoning is applied at the system level to articulate threat scenarios, design basis reliability limits, and identify critical components and overall vulnerabilities. At the micro level, emphasis is placed on execution of sound procedures, working knowledge of best practices, and perhaps most importantly, problem-solving skills and creative operator response in the face of real-time unforeseen challenges. At the micro level, security and system reliability depend on the sense-making capabilities and reaction time of system operators and occupant responders working within an unfolding event, rather than on the anticipation of planners and designers (Schulman et al. 2004).

EMERGENCY PLANNING AND THE U.S. WATER SECTOR: CURRENT STATUS

As of September 2004, the USEPA reported that all large systems had completed vulnerability assessments and certified completion of updated emergency response plans; 98% of medium-sized utilities had completed vulnerability assessments and 89% had certified completion of updated emergency response plans; and 88% of small systems had completed vulnerability assessments. Certifications of emergency response plan updates were due to the USEPA by the end of June 2004 (*Water Week*, Oct. 6, 2004).

As a key aspect of its research approach, the Stratus Consulting team conducted an in-depth survey of water utilities, including representatives from large, medium, small, and predominantly rural systems. The survey consisted of two parts: (1) a 46-question, yes/no answer questionnaire delivered and returned via e-mail; and (2) a 31-question telephone survey addressing a wide range of emergency planning topics, typically requiring approximately 1 hour to complete. Appendix A includes the questionnaires for both phases of the survey. Before they were administered, the questionnaires were pilot tested with two participating utilities and revised accordingly. Twenty-five water agencies participated in the survey, which was conduced during the spring of 2004. Major topics addressed through the survey included:

- Characterization of disaster and emergency scenarios
- Vulnerability assessment process and results
- Emergency response plan characteristics
- Emergency recovery plan characteristics
- Business continuity planning practices
- Training and drill practices

All of the utilities surveyed had an emergency response plan. More than half reported that they were in the process of updating their plans in response to perceived system vulnerabilities to human-caused events. Table 2.1 summarizes a range of survey findings, and Table 2.2 summarizes survey respondent views on the "most important" and "weakest" aspects of their utility's emergency response plan.

Table 2.1
Summary of selected utility survey responses

Top perceived vulnerabilities	Scenarios outlined in plan	Guidance used in plan development*	Frequency of training
• Wildfire	• Flooding	• USEPA	• Annual
• Insider sabotage	• Severe weather	• FEMA	• Two/year
• Seismic event	• Major chemical release	• NIMS	• Quarterly
• Contamination	• Wildfire	• SEMS	• Monthly
• Cyberattack	• Seismic event	• AWWA	• "Casual"
• Hazardous spill	• Terrorist (attack)	• None	• "Unclear"
• Structural fire	• Multiple		
• Skilled terrorist	• None		

Source: Stratus Consulting water utility survey.
*Abbreviations: FEMA, Federal Emergency Management Agency; NIMS, National Incident Management System; SEMS, Standardized Emergency Management System.

Table 2.2
"Most important" and "weakest" aspects of utility emergency plans surveyed

Most important	Weakest
• Training	• Need a "recovery" section
• Communications	• Need to address business continuity
• Staff internalization	• Need a user-friendly field version, possibly on Personal Digital Assistant (PDA)
• Staff familiarity	
• Continual updating	
• Integrated command structure	• Need more training
• Intranet availability	• Need to boil plan down to Standard Operating Procedures (SOPs), one-pagers

Source: Stratus Consulting water utility survey.

Because our survey was not designed to produce statistically valid characterizations, we are reluctant to draw "generalized" conclusions. Certain topics, however, tended to emerge frequently and/or strike an especially salient chord among survey participants:

- Roughly two-thirds of the utilities surveyed did not have facility-specific procedures.
- More than half viewed their own plan as cumbersome and difficult to implement.
- Most utilities do not differentiate sharply between "response" and "recovery" planning.
- Most utilities do not report formally modeling worst-case scenarios in an effort to gauge the need for specific emergency response equipment or supplies.
- None of the utilities reported having a completed business continuity plan, although several are working on the issue.
- None of the participants were able to articulate a clear strategy of stakeholder identification and/or role determination.
- Many utilities recognize the importance of frequent training and drills to a successful emergency response program, but training and drill activities vary substantially (see Table 2.1).
- A utility's emphasis on emergency response and recovery planning appears to be heavily influenced by its previous exposure to emergency situations.

A major theme of this report and kit of tools is that water system emergency response and recovery plans need to be well integrated with program and management support activities and training and drill regimes. In addition, the plans must be backed by an appropriate organizational culture. Utility executives, managers, staff, and partners must be mindful of security implications in their day-to-day business conduct. Our survey included several questions that yield an indication of the degree to which water systems currently supplement their plans with appropriate program backing. For example,

- Roughly two-thirds of utilities surveyed require some type of annual reassessment of their emergency response and recovery plan. The comprehensiveness and formality of these reassessments appear to vary substantially.
- Two-thirds of utilities surveyed update their plans at least annually.

- Less than two-thirds of respondents indicate that their utility conducts regular discussions of emergency response strategies with staff and management.
- Most, but not all (~75%) of utilities conduct training activities to familiarize staff, management, and executives with the emergency response and recovery plan.

GAPS IN WATER UTILITY EMERGENCY PLANNING

Interpretation of utility survey findings, literature review findings, critical review of utility emergency response plans, and input from the San Francisco project workshop were combined to reveal areas in which water utility emergency plans or planning programs tend to be weak, uneven in coverage, or fail entirely to address critical concerns. We refer to these areas as "gaps":

1. **Uneven utilization of and reliance on formal emergency scenarios:** Although some plans are clearly focused on and/or organized around one or more "likely" or "plausible" emergency scenarios, others incorporate scenarios only by means of illustration. Similarly, our literature review and utility survey both indicate that very few sources attempt to classify or characterize disaster scenarios before moving on to specify emergency management measures. We found little guidance on how to develop emergency scenarios, such as what factors to address in a scenario, how detailed to make a scenario, and whether to articulate scenarios primarily in terms of the causative event or in terms of major likely consequences (that might be caused by multiple events).
2. **Casual, informal, and generic identification of emergency response stakeholders and their differing roles:** All too often, water utility emergency response plans appear to treat stakeholder identification as an afterthought. Plans frequently include little more than a list of likely stakeholders, with little or no guidance addressing issues such as (1) how stakeholder roles might vary under different emergency scenarios and situations; (2) whether and why particular stakeholder groups should be viewed as active response partners or as passive information recipients; and (3) whether and how different stakeholders should be involved in training exercises, review of emergency planning materials, or other roles.
3. **Bringing the plan to "life":** Emergency response plans must be "wrapped in a living tissue" of program and management support, training and drill activities, and an appropriate organizational culture. Even to be minimally effective, a plan cannot be written and then merely filed—it must be shared with partners, utilized during drills and training exercises, updated frequently, and integrated across mainline business activities. The utility survey we conducted suggests that many water systems fail to conduct activities that would help to "enliven" their emergency response plans and familiarize executives, management, and staff with emergency response procedures and approaches.
4. **Emergency communications:** Many water system emergency response plans emphasize the importance of communications, but fail to provide a comprehensive approach to emergency communications. Nor do most existing plans or planning guidance resources clearly state that communication must be ongoing throughout the course of an emergency. The current state of the practice is to provide a list of helpful

tips or things to remember when dealing with the media, but step-by-step guidance is lacking. Almost all guidance materials address emergency communications, but none provide a comprehensive, step-by-step approach to help utilities develop an emergency communications plan.

5. **Business continuity and integration with core business areas:** Few utility emergency response plans address the concept of business continuity. When they do, the focus tends to be limited to information technology (IT) systems, billing, or finance. None of the industry guidance we reviewed addressed business continuity in a comprehensive, across-the-board manner. Moreover, none of the plans or guidance we reviewed outlined an approach for assessing thresholds of service disruption, priority customers, or other formalized aspects of business continuity planning.
6. **Plan usability:** Nearly half of the utilities in our survey commented that their current emergency response plan is too bulky and cumbersome for effective utilization during an actual emergency situation. Few included "rip and run" summary materials. Several survey respondents expressed a need for PDA-based templates, checklists, and plan summaries.
7. **Facility- and/or component-specific procedures:** Vulnerability assessment protocols require that utilities assess threats in terms of their potential to impact critical system components. Yet nearly two-thirds of utilities surveyed indicate that their plans are not configured to address specific components or facilities.
8. **Training:** Although many survey respondents acknowledged the importance of training, actual drill and training schedules appear to vary significantly. Roughly two-thirds of water systems conduct some type of training or orientation sessions on the plan itself, but this should be a universal practice. Actual training schedules vary from monthly to semi-annually, with some utilities training on an "as needed" basis.
9. **Balancing standard operating procedures with operator "mindfulness":** The existing portfolio of guidance materials and existing plans offer little evidence that community water systems employ practices and act to encourage an organizational culture in which executives, managers, operators, and staff are mindful of security concerns and internalize the possibility of emergency response as part of their day-to-day professional outlook.
10. **Utilization of appropriate technology:** Based on our literature review and utility survey, it is clear that water system emergency planning remains a paper-based exercise. Although many products exist, less than 15% of utilities surveyed have used software tools for developing emergency response, recovery, or business continuity plans. None of the utilities surveyed maintains a dedicated emergency response Web site (known as a "dark" site) and few appear to maintain computerized emergency notification systems. Emergency notification apparatus seems tied to old-fashioned phone trees, with no utilities reporting utilization of computer-assisted contact and information dissemination technologies.

CHAPTER 3
GUIDANCE FOR WATER UTILITY SELF-ASSESSMENT OF EMERGENCY RESPONSE PLANS

This chapter provides access to an extensive kit of tools and how-to guidance materials to assist water system staff with developing, evaluating, updating, and revising emergency response plans. It is intended to be used in conjunction with the accompanying CD-ROM, which contains the guidance materials, templates, and other resources associated with the kit of tools.

The kit of tools is structured around a framework designed to provide a step-by-step process through which utilities can periodically review and critique their emergency response plans. Illustrated in Figure 3.1, the framework depicts 12 major aspects, or "action areas," of a water utility response to a hypothetical emergency event. In other words, the framework is structured to represent a sequential logic and series of events that would likely come into play during the course of an emergency situation. The sequential logic depicted in this framework is based on our review of emergency response planning guidance and utility emergency response plans, with special consideration of USEPA (U.S. EPA 2004, pp. 103–111); AWWA (2001a, pp. 83–88); AWWA (2001b, pp. 12–15); and a major municipal utility's emergency response plan.

Although we recognize that in an actual emergency situation actions specified in this template might be performed nearly simultaneously, for the purpose of plan review and critique, we treated the actions as discrete to enhance focus and thoughtful review. We selected this framework for four major reasons:

1. The event-based logic is applicable across all types of existing plans and is not indexed to any particular format, philosophy (e.g., all hazards versus scenario-specific), or structure (e.g., USEPA versus SEMS).
2. The event-based logic meshes easily with training regimes and drill exercises, making it possible to review the emergency response plan *as part of* ongoing training and drill programs.
3. The event-based logic spans the disciplines of emergency preparedness, emergency response, business continuity, and emergency communications, helping to avoid semantic and conceptual confusion about the appropriate domain of particular planning approaches and/or response tactics.
4. The sequential nature of the framework will help water system staff and emergency response partners think through the potential consequences of events they may never have experienced.

As Figure 3.1 shows, the 12 action areas are cross-referenced (with color coding) to several widely recognized topics within the overall domain of emergency planning, including response planning, communications planning, business continuity planning, and recovery planning. As suggested previously, this is done to build conceptual and vocabulary bridges between the template presented in this document and the widest possible array of emergency planning perspectives. (For example, the California Governor's Office of Emergency Services has published *Emergency Planning Guidance—Public and Private Water Utilities*, which is organized in terms of preparedness, response, and recovery.) Applicable action areas are also cross-referenced to a topic area called "plan maintenance, security acculturation, and training." This final area is especially critical given the Bioterrorism Act's emphasis on planning as an ongoing, living activity.

In what follows, each of the 12 action areas is defined, followed by a list of questions. These questions—also available in Appendix B as a tear-out checklist—are intended to provide community water systems with a *self-assessment tool* with which to review and critique the currency, applicability, comprehensiveness, and usability of their emergency response plans and associated policies and programs. Readers will note that the self-assessment questions focus on (1) whether the plan includes and adequately addresses specific topics; and (2) whether the utility conducts related activities (e.g., drills, training exercises, plan updates, personnel performance evaluations) intended to inculcate an organizational culture predisposed to emergency preparedness and to generally envelop the plan in a tissue of supportive policy.

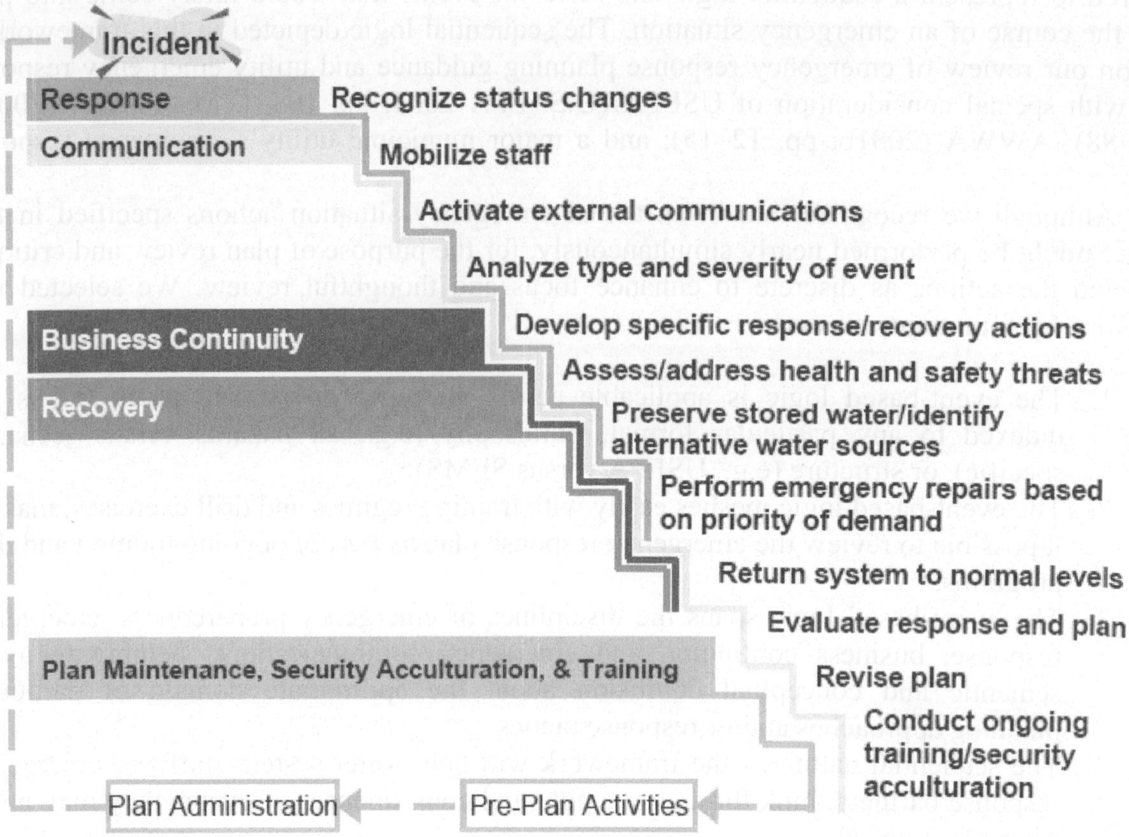

Source: Adapted from Stratus Consulting review of emergency response planning guidance and utility emergency response plans, particularly USEPA (U.S. EPA 2004, pp. 103–111); AWWA (2001a, pp. 83–88); AWWA (2001b, pp. 12–15); and a major municipal utility's emergency response plan.

Figure 3.1 Major aspects of disaster response, recovery, and business continuity planning

For each action area, the list of questions is followed by a corresponding list of resources and recommended guidance materials that are contained on the CD-ROM (see Exhibit 3.1). The resources and best practice guidance contained on the CD-ROM are numerically indexed to the self-assessment questions. If, in reviewing a particular action area, community water system officials suspect that their existing plan may be lacking or insufficient with respect to a particular question, they can immediately refer to the CD-ROM and "click through" to obtain relevant, detailed background and guidance. Guidance materials have been developed specially or excerpted from their original context to ensure focused relevance.

Exhibit 3.1
Using the accompanying
***Kit of Tools* CD-ROM**

The *Kit of Tools* CD-ROM contains resources to assist utilities in evaluating and upgrading their emergency response plans. It contains

- A list of the 12 action areas
- Self-assessment questions corresponding to each action area
- For each self-assessment question, links to recommended resources, including guidance materials, templates, and background information (in PDF format)

ACTION AREA 1—RECOGNIZE STATUS CHANGES AND ACTIVATE PLAN

A utility's emergency response plan is activated when the utility becomes aware of a threat or an incident with the potential to disrupt water service delivery or jeopardize public health and safety. This status-changing phase of emergency response, which leads to plan activation, is critical to the overall effectiveness of a water system's response to an emergency event. Utilities can recognize status changes in a variety of ways, including warnings and alerts, notification from citizen witnesses, or the utility's automated system to monitor for contaminants in the distribution system. As soon as a threat is recognized, the utility must make an *initial* assessment as to the type of threat, its credibility, and the overall severity of the incident (e.g., normal trouble, minor alert, major emergency, situation beyond the utility's control). In cases of severe emergencies and crime scenes, this step would also include notifying first responders, such as the police or fire department.

Emergency response and recovery plans must address the following questions:

A. How is a threat deemed credible?
B. Who should be contacted in case of an emergency?
C. How is an emergency reported and what information should it include?
D. How are different types of threats classified?
E. What factors determine or control activation of the emergency response and recovery plan?

The following types of tools and protocols are available on the accompanying CD-ROM for this action area:

- Guidance describing the different types of potential threats (e.g., security breach, water contamination) and how to classify them according to their level of severity

- Emergency reporting forms and notification instructions for use by utility personnel and/or staff from other agencies and organizations who may be the first to identify a threat to the water system
- Emergency reporting instructions to post at vulnerable locations and distribute to customers or the public to inform them about what to do and who to notify if they observe suspicious activity or threats to the water system
- Emergency recording form to be completed by the utility staff when they are first alerted to a threat; enables utility staff to record who notified them about the threat, along with a description of the event and immediate actions taken
- Guidelines for media access to an emergency scene

ACTION AREA 2—MOBILIZE STAFF

When a change in status has been recognized, the utility must mobilize staff to respond to the threat or incident. For some utilities, this may entail activating an emergency operations center and notifying a preassigned emergency response team (comprising, for example, the utility manager, chief chemist, communications spokesperson, and other staff). It is recognized that small systems and localities may have only one or two individuals designated as drinking water personnel. In either case, utility staff will organize themselves and initiate their emergency plan so that they will be able to respond quickly and effectively to the threat or incident.

Emergency response and recovery plans should address the following questions:

A. Has the utility established an emergency operations center or otherwise designated space for the response team to meet during a crisis?
B. Has the utility written a crisis communications plan? Does the plan include pre-scripted response information dealing with likely scenarios, answers to frequently asked questions, and summary background information on possible crisis threats?
C. Is there a pre-assigned crisis team or similar group? Do team members have defined roles and responsibilities?
D. Does the plan include an updated list of contact information for team members, top management, and potentially relevant staff and contractors?
E. Do staff and contractor listings include multiple, diverse contact modes (e.g., phone, cellular phone, e-mail, pagers, radio call numbers, home addresses)?
F. Do all applicable staff members have either full copies or applicable sub-sections/summaries of the emergency response plan?
G. Have all employees been recently briefed on the emergency response plans, emergency management authorities and responsibilities, and key emergency procedures?
H. Has the crisis team drilled or conducted exercises in which the plan has been activated and employed?
I. Has the organization designated a primary and backup spokesperson?

The following types of tools are available on the accompanying CD-ROM for this action area:

- Guidelines for developing a communication plan (which will contain many of the tools outlined in the rest of this list)
- Guidelines for establishing an emergency operations center, if appropriate
- Guidelines for defining the emergency response team
- Guidelines for defining the chain of command for responding to a threat or emergency, including how to define the roles and responsibilities of all emergency response team members
- Utility personnel contact lists providing phone numbers, pager numbers, and other contact information for all appropriate utility staff and emergency response team members
- Emergency communications messages log to record all communications and actions taken by utility personnel
- Emergency procedures review template; a brief rip and run summary of the actions to be followed during an emergency
- Instructions for activating an emergency Web site, if needed

ACTION AREA 3—ACTIVATE EXTERNAL COMMUNICATIONS

Effective communications are essential to every step in the emergency response, recovery, and business continuity process. The utility's communications staff will begin to collect information about the event (e.g., what happened, causes of the emergency, existence of any casualties) and prepare to communicate with the media, key stakeholders, and the public. The general content and format for various public notifications should be developed as part of the planning process, not during the crisis. General templates can be quickly customized according to the details of a specific situation. Note that although this framework treats external communications as a discrete action area, relevant communications actions are included under all action areas.

Emergency response and recovery plans should address the following questions:

A. Is the plan based on thorough consideration of external stakeholders? Does the analysis of key stakeholders clearly differentiate how different emergency scenarios will affect different stakeholders?
B. Does the plan include provision for external aid or mutual assistance, if required? Does the utility have an active mutual assistance arrangement with neighboring utilities or other applicable entities?
C. Does the utility maintain a regular dialogue with relevant emergency response partners and stakeholders?
D. Does the community water system involve emergency response partners (e.g., fire, law enforcement, emergency medical) in training exercises?
E. Does the plan include crisis communications lists?
F. Does the plan include crisis communication templates that help to determine who to contact first, in what order to contact key stakeholders, frequency of information updates, and pre-scripted messages keyed to particularly likely scenarios?

G. Does the plan identify a media relations officer (spokesperson), a backup media relations officer, and an emergency communications team?
H. Do all staff members know or have access to the name and contact information for the designated media relations official?
I. Have designated spokespersons received training in how to deal with the media?
J. Does the plan include a list of customers with special needs? Have steps been taken to provide ample/adequate notification and information to customers with special communication needs (e.g., non-English speaking consumers, individuals with physical disabilities, individuals with limited reading skills)?
K. Does the plan address multiple methods to deliver public messages and notifications (e.g., radio, phone banks, TV "scrollers," door-to-door canvassing, broadcast faxes, network and cable television, special Web sites)?

The following types of tools are available on the accompanying CD-ROM for this action area:

- Guidance for identifying stakeholders in the emergency (i.e., utility staff, general public, first responders, local businesses, public health organizations and facilities, media, other agencies and organizations)
- Guidance for developing mutual assistance agreements between the utility and other agencies and organizations, including a description of the necessary elements of these agreements, legal considerations, and examples (will include sample lists of agencies and organizations that have promised support, including contact information and type of assistance to be provided)
- Communications charts that list first responders, public health facilities, media, other utilities, local businesses, and other stakeholders, with contact information for each
- Guidance on notification procedures, content, format, and strategy, including examples of effective communications from the utility to the public, the media, and other stakeholders

ACTION AREA 4—ANALYZE THE TYPE AND SEVERITY OF AN EVENT

A thorough analysis of the type and severity of the emergency must be conducted. This is a more detailed evaluation than the preliminary assessment described under action area 1. It is intended to provide specific information about the damage that has occurred or may occur as a result of the incident and to characterize the severity of the emergency. The specific approach used to assess the emergency will depend on the type of threat (e.g., water contaminant, physical damage to a facility, or supervisory control and data acquisition [SCADA] intrusion).

Emergency response and recovery plans should address the following questions:

A. Does the plan include a list of critical system components?
B. Does the plan include a "damage checklist," including critical system components and likely damage modalities?
C. Does the plan include a decision tree, process schematic, worksheet, or similar tool to guide the threat evaluation process?

D. Does the plan (or an annex) contain a working list of potential contaminants, associated health and environmental impacts, and potential response/mitigation actions?
E. Does the plan include a decision tree, process schematic, or similar tool to guide analyses for potentially contaminated water samples?
F. Does the plan include a decision tree, process schematic, worksheet, or similar tool to guide site characterization activities, particularly to identify and characterize potential hazards to utility staff and other emergency responders?
G. Does the plan contain guidelines for maintaining crime scene integrity?
H. Does the plan contain—or direct responders to—water sampling guidelines and information on water contaminants, including non-traditional drinking water contaminants? Does the plan contain a sample collection, identification, and chain-of-custody form?
I. Does the plan direct emergency responders to experts, modeling tools, and other resources to help assess the flow of contaminants in the system?

The following types of tools are available on the accompanying CD-ROM for this action area:

- Preliminary damage assessment form
- Detailed threat evaluation guidance, including decision trees for evaluating different types of threats
- Guidance on monitoring the water supply, collecting water samples, and identifying contaminants
- Guidance on assessing the spread of a contaminant in the water system
- Guidance on preparing initial press releases about the incident and other communications methods

ACTION AREA 5—DEVELOP SPECIFIC RESPONSE AND RECOVERY ACTIONS

As stipulated in the Bioterrorism Act, a utility's emergency response, recovery, and business continuity plan should address event- and component-specific emergency scenarios, possibly including subplans or action plans for responding to particular types of emergencies and/or component failures. The vulnerability assessment previously completed by the utility should guide the selection of scenarios and the procedures outlined for each scenario. At the time of a threat or actual incident, the utility will reflexively implement the most appropriate action steps to address the specific characteristics, issues, and concerns raised by the event.

Emergency response and recovery plans should address the following questions:

A. Does the plan clearly articulate the water system's goals for emergency response? Are the goals listed in order of priority?
B. Does the emergency response plan articulate step-by-step processes for evaluating and then addressing the spectrum of threat scenarios that the utility is likely to face? Do response protocols in the emergency response plan differentiate between "confirmed" and "non-confirmed" incidents?

C. Is the response planning template based on the community water system's most recent vulnerability assessment or other carefully structured, fact-based inputs?
D. Does the plan include a "priority service form" or similar mechanism for ensuring that critically dependent customers receive an uninterrupted supply of water?
E. Does the plan (or annex) contain a set of response procedures for likely scenarios and events? Do response strategies focus on immediate containment or isolation of the event?
F. Do response protocols include provision to obtain and utilize alternate water supplies? If "yes," is it clear that alternative sources have been identified? Are mutual aid or other agreements already in place to authorize and initiate provision and delivery of alternative water supplies?

The following types of tools are available on the accompanying CD-ROM for this action area:

- Sample forms for summarizing existing hazard, vulnerability, and mitigation plans
- Guidelines for selecting appropriate scenarios for response/recovery action plans
- Guidelines for developing scenario-specific action plans
- Guidelines, including decision-tree tools, for reviewing and modifying action plans at the time of an emergency
- Forms to record the location of critical system information, materials, and equipment

ACTION AREA 6—ASSESS AND ADDRESS HEALTH AND SAFETY ISSUES

The most immediate actions to be taken by the utility in the face of an emergency are those that provide emergency assistance to save lives and reduce the possibility of further safety or health concerns. This includes actions to ensure public health (e.g., notifying the public of water contamination and the need to boil water, evacuating people in a flood path, ensuring that emergency medical assistance is available), protect utility personnel and functions, and repair critical water system components.

Emergency response and recovery plans should address the following questions:

A. Do response protocols clearly differentiate between known, unknown, and suspected contaminants?
B. Does the plan contain a public health response action worksheet that addresses (1) identification of the contaminant, (2) contaminant properties, (3) containment options, and (4) critical equipment within the contaminated area?
C. Does the plan (or annexes) contain summary information on properties applicable to biological (e.g., minimal infective dose, secondary transmission, susceptibility to disinfection, survivability in water) and chemical (toxicity values, aerosolization, hydrolysis, solubility, reactivity) contaminants?
D. Does the plan outline site control measures?
E. Does the plan include a description of personnel protective equipment and its location?
F. Does the plan include step-by-step decontamination procedures for likely contaminants?

G. Does the plan (or annex) contain sample notification and treatment forms, such as "water shutoff," "boil water before using," "do not drink water," and "do not use" notifications? If "yes," do these notifications (1) tell the recipient what to do, (2) describe what happened and what is being done, and (3) provide a contact source for additional information?
H. Does the plan contain a table of important translated phrases—such as "do not drink," "boil water before using," and "do not use water"—in locally applicable languages?

The following types of tools are available on the accompanying CD-ROM for this action area:

- Guidance on developing emergency response goals and procedures, including developing a public health response plan
- Guidance on how to prioritize actions to address health and safety concerns
- Guidance on procedures to protect utility personnel, records, and administrative systems to ensure that the utility can continue to respond to the emergency
- Priority service lists for the utility to use in determining the order in which to provide assistance to people and businesses with special needs
- Sample emergency notifications (e.g., to announce water shutoff, notify the public of the need to boil water)

ACTION AREA 7—PRESERVE STORED WATER AND IDENTIFY ALTERNATIVE WATER SOURCES

It is important for utilities to minimize disruption of the supply of safe drinking water during an emergency to ensure that water is delivered to their customers based on priority demand until normal operations can be restored. One way to minimize disruption is to identify alternative water supplies, including bottled water and potable water distribution points. Priority customers need to be identified and notified; they may include but are not limited to fire departments, hospitals, care facilities, and potable water distribution points. Community water systems must also take necessary steps to contain affected water supplies, preventing further spread to and contamination of other system components. In the case of a power outage, alternative energy supplies need to be available and operable.

Emergency response and recovery plans should address the following questions:

A. Does the plan (or annex) include a list of options for short term alternate supplies of drinking water, characterizing both type (e.g., bottled, bulk, untreated) and source (e.g., local government agencies, local retailers, military assets, neighboring water utilities)?
B. Is it clear that alternate supplies are substantiated through current mutual aid agreements, contractual arrangements, or other mechanisms?
C. Does the plan include information on state and federal agencies (e.g., U.S. Army Corps of Engineers, FEMA) that may be called on if local sources of water prove unavailable or insufficient?

D. Does the plan contain fact sheets and/or instructions that consumers can use to identify household sources of emergency water supply?
E. Does the plan include fact sheets or other public guidance for water purification?

The following types of tools are available on the accompanying CD-ROM for this action area:

- Guidance on establishing and implementing an alternative water supply plan to ensure safe water supply to priority customers and the community (e.g., by supplying bottled water or water from neighboring water utilities that have not been affected by the incident)
- Guidance on containing and isolating contaminated storage tanks or supply pipelines
- Guidance on conserving water and discontinuing use for all nonessential purposes (e.g., irrigation)
- Guidance on identifying and implementing alternative energy supplies (e.g., emergency generators)

ACTION AREA 8—PERFORM EMERGENCY REPAIRS BASED ON PRIORITY OF DEMAND

In addition to taking actions necessary to save lives, ensure public health, and contain the threat, the utility must also initiate emergency repairs or temporary mitigation actions to minimize additional damage. Actions should focus on restoring service to priority customers or to meet priority demands (e.g., fire departments, hospitals, care facilities, potable water distribution points). It may also be necessary to restore vital utility business functions (e.g., SCADA, financial data systems). Communication with customers should include information on estimated duration of repairs and disruption of service. Public health notifications must be reassessed and updated periodically.

Emergency response and recovery plans should address the following questions:

A. Does the plan (or annex) include a list of potential repair and reconstruction support needs, including equipment, suppliers, contractors, lifeline support, and cooperating agencies (e.g., electric utilities, wastewater utilities)?
B. Does the plan include guidance on how to dispose or otherwise manage different types of debris?
C. Does the plan (or annex) include pre-formatted work-order logs?

The following types of tools are available on the accompanying CD-ROM for this action area:

- Guidance and resources for emergency repairs (e.g., list of suppliers of spare parts and contractors to undertake necessary repairs)
- Work order logs to record repairs and the names of staff members who conduct and oversee each repair
- Guidance on tracking disruptions in operations and all damages, including associated costs incurred

ACTION AREA 9—RETURN SYSTEM TO NORMAL LEVELS

After control of the system has been regained, efforts should shift toward restoring operations to normal levels. Damaged system components should be permanently repaired or replaced. Use prohibitions and emergency restrictions on water use should be lifted only after all measures have been taken to ensure the quality and quantity of the water. Costs incurred in the response and recovery phases must be calculated and applications for compensation submitted to the appropriate federal, state, or local emergency management funds. Utility communications with stakeholders should include developing messages to explain the final stages of restoration work, address misperceptions or problems encountered during the emergency, and thank people, as appropriate.

Emergency response and recovery plans should address the following questions:

A. Does the plan include or direct users to a recovery period checklist or other guidance on how to conduct a detailed post-incident system characterization and residual risk assessment? Does this post-incident characterization process include guidance on how to coordinate with appropriate civil authorities and insurance underwriters?
B. Does the plan (or annex) include a decision tree or similar tool that addresses key restoration issues, such as long-term alternate water supplies (which may be different than short-term water supply), analysis of candidate cleanup/remediation approaches, preparation of remedial designs, and implementation of cleanup/remediation activities?
C. Does the plan contain an incident command structure for remediation activities?
D. Does the plan direct users to USEPA guidance and resources concerning remediation and recovery, including the agency's guidance for process documentation?
E. Does the plan include templates for post-event public communication?

The following types of tools are available on the accompanying CD-ROM for this action area:

- Guidance on implementing a detailed component-specific emergency recovery plan to restore normal operations
- Checklist of all actions required to restore normal operations
- Guidance for reestablishing business functions (e.g., meter reading, billing)
- Guidance on applying for federal, state, and local financial assistance to cover emergency services and restoration of normal operations

ACTION AREA 10—EVALUATE RESPONSE AND PLAN

A crucial aspect of any utility's emergency response, recovery, and business continuity plan is the comprehensive evaluation of actions taken during and after an emergency event or exercise. Until an incident actually occurs, the actions outlined in a utility's response plan are based on secondary accounts and best practices in the field, and do not necessarily reflect the realities and circumstances of each individual utility during an incident. Important questions should be answered, including: Was the response timely and effective? Which additional steps should be included in the plan to improve overall response, recovery, and business continuity?

Were customers adequately informed at all times to ensure public health and safety? Were all responders sufficiently trained and adequately prepared to deal with the incident? What could we have done better? Based on this evaluation, the utility can identify gaps in its response and its written plan (including its communication plan) and develop recommendations for improvement.

Emergency response plans should address the following questions:

A. Does the plan include forms and instructions for documenting staff activity and expenses incurred in response to a declared disaster?
B. Does the plan include sample forms and/or instructions for developing an incident report?
C. Does the plan include guidance on the preparation of a post-event evaluation report? Does guidance address issues, background, recommendations, lessons learned, and conclusions?

The following types of tools are available on the accompanying CD-ROM for this action area:

- Guidance on writing a comprehensive incident report
- Guidance on reviewing all actions taken during the incident
- Guidance on evaluating the emergency response plan and identifying gaps and shortcomings as well as overall effectiveness
- Guidance on evaluating the communication plan

ACTION AREA 11—REVISE PLAN

As we emphasize throughout this document, a water system's emergency response and recovery plan should be viewed as a living document. Based on the comprehensive evaluation process described under action area 10, the plan should be revised and updated to reflect experiences and insights gained during an incident. But the revision process must not be casual; changes must be carefully tracked and documented. The utility's communication plan should also be revised and updated. A record must be kept of all changes, and all utility staff and affiliated first responders must be informed of the changes and familiarized with new procedures. Finally, training materials must reflect all changes made, and newly revised copies need to be printed and stored at secure locations.

Emergency response and recovery plans should address the following questions:

A. Does the plan (or annex) contain a distribution list, indicating when specific individuals were provided with copies of the plan?
B. Does the plan (or annex) include a record of changes, including numbered changes, description of change, section(s) or pages affected (as applicable), date of change, and the name of the individual who authorized the change?

CHAPTER 4
SPECIAL FOCUS AREAS

Although we focus in this report on the development of a kit of tools to support and guide comprehensive review of water utility emergency planning materials, we also focus on seven broader topics critical to water utility emergency planning and response. These special focus areas tend to address issues that either support development of an emergency response plan or facilitate implementation and/or maintenance of a utility's overall security program. The seven areas described in this chapter are

1. Development and ongoing utilization of disaster and emergency scenarios
2. Stakeholder identification, characterization, and role assignment
3. Business continuity planning
4. Identification of appropriate training and drills
5. Development of communication guidance and identification of strategies and protocols for internal and external communication in the event of an emergency
6. Assessment of mutual assistance agreements and approaches
7. Classification and evaluation of software tools and applications to support and automate emergency preparedness and response planning

FOCUS AREA 1—DISASTER AND CONSEQUENCE SCENARIOS

A fair interpretation of the Bioterrorism Act and supporting statutes (e.g., SDWA) suggests that utility emergency preparedness should address a range of likely scenarios. Beyond legal and regulatory mandates are other sound reasons to structure an emergency response plan and allied training exercises in terms of specific event and consequence scenarios:

- Development and use of scenarios will help water system officials ground the planning process in specific characterizations of reality, thus helping to avoid plans that are too generic to be useful.
- The scenario development process is itself a form of training, in which utility staff members think through the implications of plausible emergency situations.

Consistent with the overall approach and perspective of this kit of tools, it is our view that the true value of scenarios lies as much with the *thinking process* that goes into their development as it does their incorporation in a paper plan. This is consistent with inputs derived through our survey of utilities, in which respondents assert that the most important aspect of their emergency response plan is "just doing it," or the process of "thinking things through."

In a scenario process, participants combine data and information with experience and assumptions to compile "stories" about plausible future events. Although fact-based, the point of a scenario is not to predict the future; rather, scenarios are heuristic devices to help people learn about the potential consequences of different events. It is true that appropriately detailed scenarios, especially if integrated with training exercises, will help utility staff rehearse specific disaster situations, thereby helping to sharpen their skills and level of preparedness. But the mindful and serious articulation of scenarios will also help utility staff recognize, work through, and respond creatively to incidents they may never have considered.

Scenarios should be based on thorough consideration of the types of disasters that can befall a water system and a characterization of the potential magnitude of associated hazards. Each utility will have its own unique list of hazards. Table 4.1 is a sample hazard summary form (AWWA 2001a). The second major aspect of a disaster scenario is a list of critical affected components. As outlined in AWWA 2001a, "[C]ritical components are those *most* vulnerable to failure or partial failure because of a disaster hazard. Failure of a critical component will reduce the system's ability to meet minimum health and safety performance goals."

The recommended way to identify critical components is to select a particular disaster characterization (and associated hazards) and then focus on those components whose failure would render the system inoperative—the critical components. Repeated application of this thought exercise will isolate the most critical components in the water system. Scenarios should also include a characterization of system performance goals, likely affected stakeholders, and service priorities. Priority demand can be defined as the minimum quantity of water needed for health and safety.

To adequately inform emergency response and recovery plan development, a scenario should include characterizations of

- The event itself, including the type of event, approximate magnitude or severity, and the timing and duration of the direct impact
- The plausible indirect impacts
- The overall affected area
- The likely water system operational implications, including critical affected system components
- The system response options (including anticipated time frame), and associated resource requirements
- The system recovery options (including anticipated time frame), and associated resource requirements
- The affected stakeholders

How many scenarios should a system consider in the development of its emergency response and recovery plan, and how many scenario-specific action plans should be included in the plan? There are no hard and fast answers to these questions. Our survey of utilities indicates substantial divergence in the use of scenarios. Some utilities articulated only a single, most likely scenario. On the other extreme, one utility included short, 3- to 4-page scenarios for almost 20 different potential emergencies. In its guidance to water utilities for addressing human-caused and technological emergencies, the USEPA includes five illustrative scenarios, along with concise action and response summaries (U.S. EPA 2002).

Table 4.1
Sample hazard summary*

Type of hazard	Estimated probability	Estimated magnitude	Comments
Earthquake	1 in 60 years	7.0 (Richter scale)	
Landslide	Medium–high		In slopes of ≥30%
Tsunami and seiche	None		
Hurricane	None		
Tornadoes	Low		
Floods	Low–medium	100-year flood to elevation = 1,020 ft	At treatment plant
Forest or brush fires	High		Dry creek watershed
Volcanic eruptions	1 in 300 years	150 miles away	Mount Nueces
Other severe weather			
Snow or ice	None		
Extreme heat	High	100-year drought	Reservoirs depleted
Wind	Medium	60–80 mph	Usually in winter
Lightning	Low		
Other			
Waterborne diseases	Low		Cryptosporidiosis
Hazardous material release			
Chlorine	Medium–high	1-ton containers	Earthquake damage
Other spill	Medium	Tanker car	Dry creek reservoir
Structure fires	Low		
Construction accidents	Medium	Line damage	In older area of system
Transportation accidents			
Road	Low		
Rail	Medium		Railyard near warehouse
Water	Low		
Air	Low		
Nuclear power plant accidents	Low	Contamination	Lake West reservoir
Nuclear bomb explosion	Low		
Vandalism, terrorism	Medium		Storage tanks
Riots	Low		
Strikes	Low		

Source: Adapted from AWWA (2001a).
*Note: This is a sample hazard summary form. The form can be adapted to meet individual system needs.

To be clear, we view scenarios as the developmental building blocks for emergency response and recovery plans. Plans should be based on careful consideration of alternative scenarios. Following Grigg (2002), we suggest that systems develop and populate a two-by-two matrix for disasters. Illustrated in Exhibit 4.1, the matrix should include high-impact, low-probability incidents; high-impact, high-probability incidents; low-impact, low-probability events; and low-impact, high-probability events. Although a broad range of disaster scenarios, both human-caused and natural, should be considered in the development of the plan and in training activities, the plan itself should include specific action subplans only for a small number (probably eight or less) of the most likely, highest impact scenarios. Note, however, that as suggested under focus area 4, a utility's training regime should include a mix of likely and unlikely scenarios.

Information on scenario development is contained in action areas 1A, 5B, 5E, and 12.

Exhibit 4.1
Scenario matrix for disasters

High-impact, low-probability incidents	High-impact, high-probability incidents
Bioterrorism in plant or distribution system Failure of chemical storage tanks Severe and long-lasting shortages Attacks on plants Extreme natural disasters (earthquake, flood, hurricane, firestorm, and others)	
Low-impact, low-probability incidents	**Low-impact, high-probability incidents**
Chemical in raw water reservoir	Frequent natural disasters Health problems and contamination Power outages Mechanical breakdowns Main breaks System upsets and backflow incidents Nuisance break-ins and pranks HAZMAT incidents Droughts of short duration Cyberattacks Weather events, especially freezing Contractor damage Network and SCADA failures

Source: Grigg 2002.

FOCUS AREA 2—STAKEHOLDERS

In the context of water supply, stakeholder groups are parties affected by, or perceiving themselves to be affected by, a utility action, including a utility's response to an emergency situation. The term "parties" includes organizations, groups, and even small aggregations of individuals. Parties may be aware or unaware of how utility actions affect them; they may be actively involved in an emergency situation (attentive) or largely indifferent (inattentive).

As described by Babiuch and Farhar (1994), stakeholder analysis is a formal process through which to identify if and how various parties are likely to be affected by a specific event or action. Impacts are defined as changes in health or well-being brought on by an event and/or a utility's response to the event. Impacts may be either direct or indirect, and depending on the event, are likely to affect different stakeholders in different ways and to differing degrees. Stakeholder groups must be identified systematically to avoid inadvertently overlooking affected parties or to overemphasize the needs of relatively unaffected parties. In addition to identifying stakeholders by contemplating the likely impacts of a particular emergency scenario, utilities can also identify likely stakeholders by considering the plausible "domains of influence" associated with a specific emergency scenario or response approach. Domains of influence can be characterized by listing the functions that a particular scenario might affect. Functions are those social, economic, and/or environmental activities on which the contemplated scenario (or the utility's planned response approach) might have an impact.

Utility relationships with stakeholders typically reside along a continuum between "reactive" and "proactive." In the case of a water system's response to an emergency event, our research strongly suggests that a proactive approach and posture is much more effective than a reactive approach. Reactive utilities are continually playing "catch-up," and run the risk of allowing small-scale events to "spin" out of control and be viewed as major organizational crises (Pearson and Clair 1998). In the context of emergency response planning, it is important that community water systems and other water utilities carefully consider which parties will be stakeholders under alternative emergency scenarios, how different response strategies will affect different stakeholders, whether impacts on specific system facilities and/or components will differentially affect different stakeholders, and how the roles of different stakeholders may vary from one situation to another.

It is therefore insufficient to merely include a list of likely stakeholders in an emergency plan. A thorough consideration of stakeholders should include three basic principles:

- The impacts of an anticipated emergency event, or event scenario, should be considered systematically. Formal consideration of each event on each group of likely stakeholders will enable water systems to better anticipate and prepare for emergency impacts across the entire stakeholder population.
- Stakeholder groups should be identified comprehensively. Different emergency scenarios will affect different parties, or at least have an impact on different parties to differing degrees. The process of identifying affected stakeholders must be informed by specific emergency scenarios.
- Analysis of stakeholder impacts to alternative disaster or emergency situations should be considered when the utility is formulating its response protocols and action plans for specific facilities or components.

Resources that can assist utilities in addressing stakeholder issues can be found action areas 1B, 2B, 2C, 2D, 2E, 3A-K, and 12B in the kit of tools.

FOCUS AREA 3—BUSINESS CONTINUITY PLANNING

AwwaRF Project No. 2929 called for evaluation and review of the topic of "business continuity planning" (BCP). Accordingly, the Stratus Consulting research team sought background information on the BCP concept, and explored its application within the context of the water sector. Input on BCP was derived through targeted literature review, our utility survey, and focused discussion at the San Francisco Workshop.

Increasingly, modern utilities operate in a seamless environment with their customers. Disruption of the water supply may immediately and directly affect a customer's ability to continue operations or provide a critical service. As understood by many, BCP is a recognition of the importance for planning utility operations into and through an emergency event (Hopkins 2004). As defined by Moore (2000), business continuity planning is "the documentation of . . . strategies, procedures, [and] resources . . . utilized by an organization to respond to, recover from, resume and continue operations in the event of a substantial disruptive incident."

Business continuity planning places a decided emphasis on strategies to minimize, or perhaps even eliminate, service disruptions, at least with regard to specific priority customers. Business continuity planning, then, might best be understood as emergency recovery planning with a broader scope (Kirvan 2003). Considered in this way, it can be argued that BCP does not differ in any fundamental sense from the thorough and effective application of traditional emergency planning models. As Exhibit 4.2 indicates, BCP shares a common domain of activity with disaster recovery and restoration planning.

Figure 3.1 shows that the tools and approaches associated with BCP tend to overlap with those utilized in typical water utility vulnerability assessment approaches and emergency response plans. However, BCP differs from conventional approaches to emergency planning in two respects: (1) it places a more explicit emphasis on the characterization and estimation of the impact potential associated with service disruption scenarios, and (2) it adheres to specific tolerance thresholds for service interruption. In its extreme formulation, BCP could involve an emergency recovery regime with no tolerance for disruption, at least with regard to specified priority customers. As Smith (2003) observes, business continuity planning is not merely a set of tools, techniques, and mechanisms, but rather adoption of more reflective and proactive attitudes with regard to emergency response and recovery planning.

Exhibit 4.2
Common elements of business continuity planning

- Emergency teams and key personnel
- Team responsibilities
- Risk analysis results
- Disaster scenarios definition
- Business impact assessment
- Inventory of critical assets
- Recovery time-frames
- Prioritization of recovery activities
- Business process reactivation activities
- Disaster recovery strategies
- Awareness and training activities
- Response, recovery, and restoration activities
- Plan documentation
- Plan distribution
- Plan maintenance

Source: Data from Contingency Planning and Management (CPM) 2003.

The BCP emphasis on uninterrupted service delivery is consistent with a key theme found in the literature on so-called high-reliability organizations (HROs). Some enterprises are inherently risky, involving complex systems, hostile and unpredictable operating environments, unrelenting public scrutiny, and potentially catastrophic consequences in the event of a failure. The day-to-day operations of aircraft carrier flight decks, nuclear power plants, wildfire fighting crews, and air traffic control crews can be characterized by technological complexity, challenging environments, and the potential for high-visibility and high-consequence failures. Yet organizational experts and disaster researchers have discovered that enterprises such as these tend to have less than their "fair share" of accidents and failures. To find out why, we can look at the operational characteristics that distinguish HROs:

- A top-to-bottom assumption that threats and consequently errors are omnipresent and that unrelenting vigilance is the price of success
- A parallel assumption that sources of error and risk are dynamic, not static, so that monitoring mechanisms and protocols themselves must be constantly renewed and reinvigorated
- A strong belief that the operating environment is a constant source of threat, requiring constant vigilance, even when things seem to be going well
- Maintenance of redundant modes of problem solving at the operational level, and resistance to pressure to resolve or "rationalize" the process by adopting a best practices approach

- The creation, maintenance, and exercise of multiple simultaneous informal organizational structures adapted to contingencies
- A relative empowerment of organizational units and activities dedicated to searching for incipient, latent, or potential failure
- A marked reluctance to rely on formalized regulations and codes ("going by the book"); a need to augment plans with ceaseless testing, practice, and experiential learning
- Acceptance of the proposition that it is impossible to adequately characterize all possible risk scenarios; a need to constantly imagine new sources of risk

In short, HROs are enterprises that view risk as a normal operational variable that must be constantly articulated and dynamically managed. Emergency planning, vulnerability assessment, and other risk management activities, then, are not viewed as static, one-time activities or even as activities that can be adequately addressed through periodic exercises, but as salient aspects of day-to-day operation.

There are clear differences between the operation of a typical water utility and that of, for example, an aircraft carrier. Although it may be unreasonable to posit a far-reaching and robust analogy between HROs and water utilities, it does seem that the overall HRO outlook is applicable to emergency planning in the water sector, especially with regard to the implementation of an aggressive approach to business continuity planning.

Information that most closely relates to business continuity planning is found in action areas 5-9 in the kit of tools, especially in action areas 7 and 8, which begin to address priority of demand issues.

FOCUS AREA 4—TRAINING AND DRILLS

During the literature review and utility survey portions of our research, we focused heavily on characterizing water utility emergency response training activities, including critical review of actual exercises and drills. Introduced under focus area 3 (Business Continuity Planning), the HRO literature base was a fruitful resource for finding training techniques that can help utility executives, managers, and operational staff assume a more effective and proactive role in recognizing and responding to unexpected, atypical, and potentially extreme events such as terrorist attacks.

Emergency response training is conducted to (1) familiarize employees and selected stakeholders with emergency procedures and the content and structure of the emergency response plan; (2) train employees (and selected stakeholders) on how to implement the plan; (3) evaluate the plan to ensure that it is effective *and* usable under stressful conditions (see chapter 5 for a thorough discussion of plan usability); (4) identify areas of plan implementation that may require more, or more elaborate, training; and (5) to inculcate employees, managers, and executives with a mindful and creative approach to emergency problem solving.

Emergency response experts recognize four basic types of training exercise, described in the sections that follow. These four types can also be viewed as incremental stages in a multilevel training program.

Orientation Sessions

Orientation sessions are configured to introduce new or updated information, ideas, procedures, or tools to individuals expected to carry out specific procedures or to a new employee. These sessions include basic instruction and explanation of emergency procedures along with clarification and highlighting of critical plan elements. They can be offered in various formats (e.g., seminar, monthly bulletin, periodic lectures). Orientation sessions can be characterized by low stress, relatively short duration (1-8 hours), low cost, and high frequency (many sessions can be conducted on an annual basis as refresher training for existing employees and as training for new employees).

Tabletop Exercises

So-called tabletop exercises present students with an emergency scenario in a sequence of steps or as a single event. Alternatively, participants can select a specific event scenario. Participants respond verbally to a series of questions about the scenario. Tabletop sessions can also include role playing, in which participants act out critical steps, recognize difficulties, and resolve problems in a nonthreatening environment. Responses are then evaluated to see if they are compatible with the plan. Revisions to the plan and/or training sessions are made, as needed.

Tabletop sessions are characterized by relatively short duration (3-8 hours) and medium-high frequency (a utility and its partners can plan and conduct several per year), and require moderate preparation time (a few weeks).

Functional Exercises

In a functional exercise, a "training team" sends a series of pre-scripted messages to exercise participants. Participants respond to these messages by simulating the actual procedures they would utilize in response to the event. Functional exercises can involve mobilizing participants to other sites in an attempt to complete specified actions, establish communications, and coordinate with other team members as required in the plan. Training exercises can involve activities such as running emergency generators, practicing media interviews, or isolating parts of the system.

Functional exercises require moderate duration and a significant investment in planning and preparation time (up to 6 months). Depending on available resources, it may be impractical to conduct functional exercises more than once or twice a year. They do, however, provide a useful and effective way to test the emergency plan.

Full-Scale Drills

Full-scale drills involve the mobilization of field crews (including the incident commander) and equipment to conduct mock responses to a simulated emergency. The problem scenario is presented to the participants and they respond as directed by the emergency plan. Full-scale drills are the most costly and time-consuming of the different types of training approaches. In most jurisdictions, full-scale drills are only conducted once every 4 or 5 years; and may require as much as 6 months to 1 year for planning and preparation.

Detailed information on emergency training exercises is contained in action area 12A-G of the kit of tools.

The Need to Combine Training With Plan Evaluation and Security Acculturation

Whatever combination of training approaches is used, it is important to use training exercises as a tool to critically evaluate the existing emergency plan (assess content and usability of the plan; identify weaknesses, gaps, other problems). At the conclusion of a training exercise, it is recommended that participants and the training implementation team meet to debrief and discuss whether training objectives were met, whether test results are valid, what problems were encountered during the exercise and what corrective actions can and should be taken, whether gaps exist between the plan and the actual test results, and perhaps most importantly, whether and how the plan should be modified.

As emphasized earlier in this report, one of the most insidious aspects of terrorism is its unpredictability. How does a utility train for an event that it cannot even conceive? It is not possible to develop training exercises that address every disaster that could possibly befall a public water system. But there are ways in which a utility can think broadly about the full range of problems it might encounter and develop preparedness strategies.

As described by Mitroff and Alpaslan (2003), some organizations prepare effectively for unanticipated crises by thinking about them at random. The carefully orchestrated use of a random-selection technique forces utility personnel and other training partners to abandon conventional modes of thought, stretch their frames of reference, and envision abnormal events and viable solutions. One tool that can be used to compel utility staff to think randomly is called a "wheel of crisis." Shown in Figure 4.1, the wheel lists all the types of emergencies that the utility staff can imagine. Staff spin the wheel, roulette style, and then discuss emergency response steps and approaches for each type of disaster. A similar technique involves the use of "internal assassins," or small groups of utility staff asked to imagine themselves as terrorists bent on attacking the water system. Utilizing their intimate knowledge of the system, these groups devise ways to damage the system, introduce contaminants, or otherwise threaten public health and well-being. The schemes hatched by the internal assassins can then be used as training exercises for other utility staff.

Water utility managers and executives should recognize that there is an inherent tension involved in effective emergency response training programs. On the one hand, it is essential that utility staff and their partners become thoroughly versed and comfortable with basic emergency response techniques, procedures, and plans. As the old adage has it, "stress makes people stupid," and the onset of an emergency is not the time to first open the emergency response plan. On the other hand, it is important that emergency response training does not become rote. In other words, training programs and other aspects of utility culture must be designed and managed to make sure that personnel at all levels think creatively about potential emergencies, system vulnerabilities, and effective response strategies.

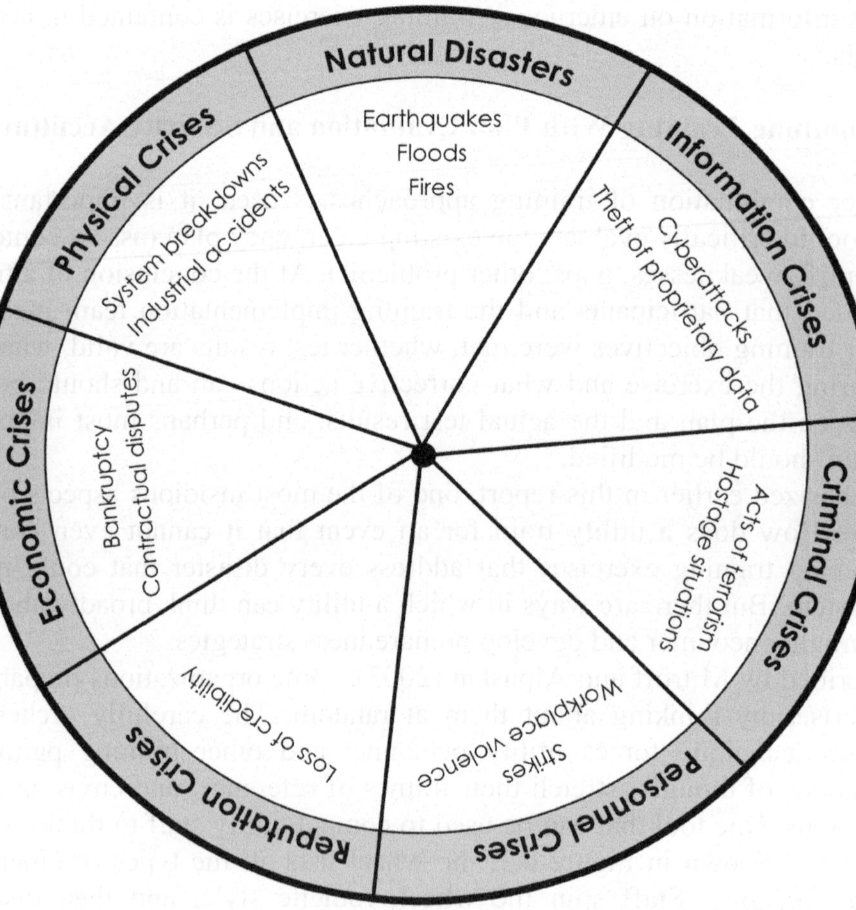

Source: Adapted from Mitroff and Alpaslan (2003).

Figure 4.1 The wheel of crisis

FOCUS AREA 5—INTERNAL AND EXTERNAL COMMUNICATIONS

In addition to the overall literature review and utility survey, communications professionals on the Stratus Consulting research team conducted a focused critique of 22 emergency response plans (from water utilities and a variety of other organizations) to assess the adequacy of their approach to crisis communications.

Although the review uncovered ample rhetoric on the importance of a communication plan and/or effective communications practices, we found little in the way of specific step-by-step guidance. Utility guidance also suffers from strategic problems, such as the following:

- Inadequate measures to sensitize water system staff to the dynamics of the "media frenzy" that typically accompanies a crisis event
- Lack of emphasis on the need to utilize the media proactively to communicate with the public

- Overemphasis on phone trees and failure to identify other channels for communication and information dissemination
- Little emphasis on the usability of emergency response materials
- Inadequate background on how relatively minor events can "morph" into major organizational crises through inadequate attention to public and media communications

Given the nature of the terrorist threat (including a greater inherent potential for panic than with other types of disasters), it is essential that utilities establish a comprehensive communication program, including clear assignment of individuals authorized to speak with stakeholders and the media, well in advance of any actual disaster, disruption, or contamination event. Also essential are pre-established protocols for communication *within* the utility, and with other emergency response entities (e.g., first responders). For both internal and external uses, a water utility emergency response plan should be configured to provide an orderly and systematic approach for managing communications during a disaster or crisis situation. It must serve as an immediate and transparent reference for management, employees, cooperating first responders, and clients who may be required to respond during a disaster situation.

There is necessarily overlap between the utility's communications plan and its emergency response plan. In fact, as seen in the kit of tools, communications tasks are an integral part of a utility's response to an emergency, occurring in every step of the emergency response process. Because a clear and well-thought-out communications strategy is essential to successfully responding to an emergency, however, it is our recommendation that water system emergency response plans include a discrete emergency communications subplan, annex, or component.

Important components of a crisis communications plan include the following:

- A clear demarcation and procedures for addressing the three time phases of a crisis: planning before a crisis, management of the crisis during the event, and management of the situation after the event
- Development of a crisis communications response team
- Selection of and media training for designated spokespersons
- Creation of a wide range of news releases and message points in advance
- Development of systems and procedures to disseminate information internally, to the media, and to stakeholders and the community
- Simulation or role-playing exercises to help spokespersons anticipate questions, responses, and problems

Stratus Consulting has developed an emergency communications plan outline, which is included in action area 2B of the kit of tools. The outline contains links to relevant communications tools contained within the kit of tools action areas.

FOCUS AREA 6—MUTUAL ASSISTANCE AGREEMENTS

During a disaster or emergency event, a water system may need assistance from outside parties in responding, recovering, or returning to normal operations. Such assistance might involve borrowing heavy equipment, pipe, chemical supplies, skilled physical labor, generators, analytical services, or technical expertise. Typical sources of assistance to a water supplier are other water utilities with the expertise and capacity to provide assistance when it is needed; however, a variety of public and private organizations can also furnish assistance, including state or county transportation departments, municipalities, electric utilities, construction companies, and military installations. Emergency assistance is frequently formalized through mutual aid agreements.

Mutual aid agreements exist in many forms and may be regional, statewide, or localized. Depending on the functional needs of the parties to an agreement, they may involve contiguous jurisdictions or dispersed entities. Sometimes mutual aid agreements fall within the purview of a covering administrative or statutory regime; other times they are simple, contract-like arrangements between two or more organizations. In some cases, mutual assistance agreements are augmented by dedicated support staff, budgetary authority or other funding provision, and supporting resources such as Web-based equipment inventories, joint emergency response plans, and/or combined incident command systems. Some mutual assistance agreements are omnibus, involving a broad range of circumstances and services; in other instances they are relatively narrow and may involve only a single resource or service. Most mutual aid agreements involve a formalized, contractual arrangement, but there are examples of successful, long-term mutual aid based on unwritten "handshake" agreements.

As a key aspect of our research and best practices evaluation, the Stratus Consulting team characterized the role of preplanned regional collaboration in response and recovery to disasters. Specifically, we developed a multiattribute framework to formally characterize and compare various agreements. Based on this framework, we explored the long-term value, inherent efficiency, benefits, and possible problems associated with formal agreements for delegating roles and defining support that can be relied on from neighboring water utilities. In addition to a literature review and analysis of applicable utility survey input and review of 12 existing water utility mutual assistance agreements, we conducted a structured roundtable discussion with the four major Washington, D.C.-area water utilities and the Metropolitan Washington COG. Focusing on mutual assistance, the roundtable with DC-area utilities was critical because it gave us insight into the successful operation of an informal approach to mutual assistance.

Based on our review of existing mutual assistance arrangements, the practice tends to adhere to the following set of guiding principles:

- Participation must be strictly voluntary; agreements cannot create a "duty" to respond.
- It must be clear that potential lenders cannot be held liable for a failure to lend assistance.
- Resources can be withdrawn at any time, even if they have already been lent.
- All equipment is provided "as is," with no representations or warranties as to its fitness for a particular use.
- Use, repair, and/or replacement costs are borne by the borrower.
- Employees must be willing to participate.
- Incident command is the responsibility of the borrower.
- Borrowers must be willing to comply with the lender's statutory requirements (if any).
- Inventory information needs to be shared, participants should develop a list of priority materials and personnel, and the lists should be updated frequently and shared among all participants.

- Although response is not mandatory, assistance requests should take priority over routine matters.
- Parties to an agreement should establish a hierarchy of needs to help guide lenders with more than one request.
- Parties to an agreement should develop mutual aid checklists for borrowers, lenders, and incident commanders, including step-by-step procedures on how to request aid.
- Mutual aid agreements should be couched in terms of (a) emergency response plans, and (b) lists of emergency contacts.
- Mutual response agreements (and supporting inventories) should be reviewed and updated at least annually.
- Mutual assistance agreements should include a mission statement.
- Costs and reimbursement are frequently a point of contention in the arena of mutual assistance, with conflicts fairly common. It is important that participants (a) spell out detailed cost accounting and reimbursement arrangements, (b) work with state or regional government units to establish an independent budgetary authority, or (c) rely on trust and existing handshake arrangements, accepting the attendant risk of financial dispute.

Examples of mutual assistance agreements and other resources for developing mutual assistance agreements are contained in action areas 3B, 5F, and 7B of the kit of tools.

FOCUS AREA 7—SOFTWARE TOOLS AND APPLICATIONS

In recent years, water utilities have been confronted by a diverse array of commercial off-the-shelf and custom software products designed to aide in disaster preparation, disaster response, disaster recovery, and emergency management. These products range from generic software packages developed to support the creation and analysis of fault trees to specialized and industry-specific applications.

The Stratus Consulting research team canvassed available product offerings and developed an inventory of emergency planning and disaster management software applications available to the water industry, including applications that are Web-enabled. With 32 products, the inventory includes a description of each product, a listing of key capabilities and parameters, and a brief assessment of each product's usability and applicability in the context of water utilities.

Overall, water systems can choose products from among five broad, somewhat overlapping categories of software tools and applications:

1. **Vulnerability assessment applications:** These tools enable a water system to input perceived threat information and data on system assets to produce utility-specific risk factors and cost-justified remediation plans. Although outside the scope of this analysis, the vulnerability assessment process is mandated by the USEPA and provides key inputs to the emergency planning process.
2. **Incident control and damage assessment applications:** These tools offer a formalized structure and prompts to help water system personnel plan and implement emergency response activities. Typical applications include a database of critical

system assets and a template for coordinating the tasks and activities of an emergency response team.

3. **Disaster recovery and restoration planning tools:** These applications utilize a template structure, drop-down menus, and similar guidance mechanisms to help utilities produce disaster recovery and emergency response plans. Applications store team roles, responsibilities, critical assets, procedures, and equipment inventories. Some applications of this type can be used to generate prioritized risk-based scenarios and response action plans based on data inputs.
4. **Water contamination analysis tools:** These applications utilize distribution system fate and behavior models to characterize and predict the flow of a contaminant within the water system. They can be used to help assess the consequences of transport of a contaminant within the water distribution system; assist in the identification and location of a contaminant; characterize the population at risk from a specific contaminant; and, in some cases, conduct "what if" analyses for alternative response and remediation strategies.
5. **Emergency communication and notification products:** These products are designed to facilitate an organization's response to an emergency event, mostly through enhanced notification of utility staff, stakeholders, emergency responders, and customers with special needs. These systems store contact information and utilize multiple media (e.g., phone, fax, e-mail, pagers) to quickly and simultaneously alert and instruct recipients about an emergency event. These applications store earlier communications and generate communications logs, and in some cases, store pre-drafted messages or provide templates for certain types of emergency messages.

In almost all cases, the application walks the client through a structured question-and-answer process tied to vulnerability calculators and/or pre-defined response plan text. In some cases, the tool enables the user to customize the interrogatory process and the pre-set associations to better fit the individual utility's system design, components, or operating environment. As the client input/interrogatory process is completed, information is automatically associated with the appropriate plan areas and integrated into the draft plan, policies, or procedures.

In general, strengths and weaknesses associated with these tools include the following:

Strengths

- By their very nature, these systems provide increased availability of data critical to emergency response planning; ability to call up and compare current and historical data; and generally easier data maintenance and storage.
- Over the long term, conversion from a manual to an automated process will likely save time and reduce costs for the utility.
- Planning templates, drop-down menus, and interactive checklists may help to ensure that planning efforts are relatively more comprehensive and do not overlook critical steps or aspects.

Weaknesses

- These systems require high initial costs for implementation, including data capture, data entry, and user training (especially for customization).
- Generic products may surprise users with extensive need for customization.
- No single product offers a complete emergency response and recovery package; none of the products reviewed helps to extend emergency response into the broader realm of business continuity planning.
- Automated processes have the potential to seduce users into a relatively lax mindset with respect to emergency planning; similarly, emphasis on pre-scripted scenarios may lull users into a false sense of readiness, leaving them unprepared to deal with unanticipated events.
- Software needs to be used, data entered, and information customized, if necessary, prior to an emergency. If these steps are not taken before an emergency, it will be unreasonable to expect to use the software in a timely and effective manner during an emergency.

Although some of the systems seem clearly superior to others, none of the reviewed applications should be viewed as a "magic bullet" for all of a water system's emergency response and recovery planning needs. The software evaluation report, including a description of our evaluation methodology and a full summary of all products reviewed, is included as a resource document in the kit of tools (see action areas 2B, 3F, 4B, 4E, 4I, 5B, 5C, and 5E).

CHAPTER 5
EMERGENCY RESPONSE PLAN CONFIGURATION AND STRUCTURE

This chapter presents our recommendations for how to configure and structure an emergency response plan. It also includes suggestions for ensuring that plans are designed so they can be used easily and efficiently under high-stress emergency conditions.

RECOMMENDED PLAN STRUCTURE

As mentioned in chapter 2, most water utilities subject to Section 1433 of the Bioterrorism Act have already certified completion of updated emergency response and recovery plans to the USEPA. Survey research and literature reviews conducted in support of this project indicate that existing water utility emergency response plans vary significantly in terms of scope, structure, length, topics addressed, specificity of scenarios, and other formative and substantive factors. Utilities draw on a wide range of guidance materials and other resources to develop and maintain their emergency response plans and programs, including the USEPA, the Department of Homeland Security (DHS), FEMA, water sector research organizations, the American Red Cross, various state emergency management agencies (especially in California), and others. In other words, the existing water sector emergency response and preparedness guidance portfolio is highly diverse and tailored to account for the unique circumstances and capabilities of individual utilities.

For utilities contemplating major change or those that have yet to complete an emergency response plan, however, we recommend adoption of the generic planning frameworks published by the USEPA. Adoption of the agency's framework was strongly advocated by utility representatives and water sector stakeholder participants at the project workshop, citing reasons such as the following:

- The USEPA exercises statutory oversight for the SDWA, relevant provisions of the Bioterrorism Act, and other emergency planning mandates for the water sector.
- Adoption of the USEPA structure will help to foster a common parlance among utilities and exchange of best practices information.
- Adoption of the agency's structure and parlance will facilitate development of mutual assistance agreements, joint training exercises, and other inter-utility resources and activities.
- The USEPA guidance is scaled for small, medium, and large systems, and is easily accessible via www.epa.gov/safewater.

As outlined by the USEPA, an emergency response plan should be organized into a relatively small number of logical sections (eight or less), so that information is easy to find and the overall structure possible to remember, even under stressful conditions. Appendices should be used for supplemental materials such as site and system schematics, event- and scenario-specific action plans, prewritten press releases, and various checklists. A suggested plan configuration follows.

- **Section 1—Introduction:** Describes the purpose, goals, access control protocols, and overall organization of the emergency response plan.
- **Section 2—General Emergency Planning Information:** Describes emergency planning partnerships, mutual aid agreements, and the relationship between the emergency response plan and the water system's other emergency response policies, procedures, programs, and documents.
- **Section 3—Emergency Response Plan—Core Elements:** Provides specific information about the system, describes emergency response roles and responsibilities and incident command information, and describes communication procedures, personnel safety protocols and provisions, and alternative water supplies. This section also presents information on emergency equipment and chemical supplies, property protection, and distribution system sampling and monitoring, along with any necessary procedures.
- **Section 4—Decision Process and Emergency Plan Activation:** Explains how threats may be received by the community water system and outlines steps to take in activating the emergency response plan and/or emergency operations center.
- **Section 5—Emergency Response, Restoration, and Recovery:** Describes three phases of an emergency, with general actions and guidance for each phase. These procedures should be used in conjunction with scenario-specific action plans in an appendix to the emergency response plan.
- **Section 6—Action Plans:** Presents a summary of the scenario- or incident-specific action plans that the system has created to guide response to and recovery from specific threats. The action plans themselves will be organized in an appendix to the plan.
- **Section 7—Emergency Plan Approval, Update, and Training:** Describes the emergency response plan review, approval, and update processes; discusses how the utility evaluates the effectiveness of the plan and supporting procedures; and contains information on the system's emergency training program.
- **Section 8—Additional Resources and Appendices:** Outlines references and gives Web links to useful background information, data, and other emergency planning materials useful to water utilities. Appendices also contain specific action plans, forms, worksheets, pre-scripted and template press releases, mutual assistance resource lists, and other information and resources that may be used in an emergency response situation.

COMMUNICATIONS AND DECISION MAKING UNDER STRESS

Our research for this project also focused on the specialized areas of high-stress cognition and risk communications, resulting in a body of information and findings highly relevant to individual and group response to disasters and other stress-inducing extreme events. Disasters and other emergency events will likely present utility staff, managers, executives, and other responders with a set of problems that may involve time pressure, uncertainty, and high stakes—three factors that combine to create stress and make it difficult to implement emergency response procedures (Baddeley 1992, Newell and Simon 1972, Lavie 1995).

All humans have limited resources for cognitive processes. A generally accepted model of human cognition divides memory into two categories, long-term and short-term (or "working") memory. The limitations of working memory are well documented. For instance, it has been demonstrated that most people can remember and attend to between five and eight things at any given moment. Another limitation is that humans are serial information processors. We have inherent difficulty dividing attention among several tasks, or attending to all the information our senses take in. These limitations in attention and memory interact with our perceptual processes. As a result, we often do not perceive most of the information that is available to us, a factor that is exacerbated by the stress of time pressure.

Because of such factors, designing emergency response and decision support tools so that they are operable during stressful situations and so that they do not increase the cognitive effort required to respond to a perceived emergency is essential (Tufte 2001). Practices that will aid in this context include the following:

- **Simultaneously Display All Relevant Information:** This means that the emergency response plan and each major section or aspect of the plan should be reduced to a single page (or single view in the case of an electronic or Web-based document) summary of the plan and/or section as a whole. Clearly, such a summary cannot contain detail that is requisite of the plan as a whole, but the visible items are intended to serve only as reminders or memory prompts. It is also important that each subsection include a series of icons representing the plan as a whole, helping the responder to avoid losing track of the overall situation.
- **Sequentially Order Relevant Information:** The structure of most emergency plans reviewed for this project does not reflect the temporal sequence of events likely to unfold in an emergency situation. The single-page summaries described in the last bullet should be arranged in sequential order, perhaps utilizing a sequential framework such as that adapted for this project. This too will help to reinforce memory, and also prompt response team members to anticipate next actions.
- **Use Small Multiples Whenever Possible:** The action-relevant sections of the plan should be kept to eight or less to accommodate the number of things we can attend to at a given moment. Using such a structure to begin with makes it easier to develop rip and run summaries of the plan and key annex materials.
- **Use a Checklist Format to Aid Memory:** Responding effectively to fast-paced events may require actions to be carefully timed. Because the stress of time pressure can impair memory, it can be difficult to remember which actions have already taken place. Rip and run summary sheets should include check-off boxes to show that an activity is complete.

USABILITY "STRESS TESTING"

Well-established methods and procedures are available for assessing the usability of paper reports, Web sites, and electronic information applications. These techniques help developers ensure that a report or site is logical, intuitive, and transparent to target users. Usability testing involves systematic review of an information product according to a set of common usability principles that address form, overall structure, and content. Examples of usability principles include navigability (i.e., users should always know where they are in a report or Web site and where else they can go), consistency (i.e., users should not have to wonder whether different words or actions mean the same thing), and accessibility to users of varying skills and capabilities. Usability testing is especially critical for action plans, Web sites, and other applications intended to be used under emergency or extreme conditions.

It is our recommendation that utilities subject draft emergency response plans to a testing regime that simulates typical users and use-cases. Test participants can be asked to perform a variety of tasks while an observer records what the user says and does, and more importantly, whether the user successfully completes the task. A variety of types of tasks can be performed during the test, including timed, information-gathering "scavenger hunts"; interpretation of key instructions; ability to recall key aspects of the plan structure and content; and "stress tests," in which participants begin on random pages and are asked to explain their relationship to the plan as a whole.

Plan usability testing sessions can be conducted much as one would conduct a tabletop training exercise (see chapter 4, focus area 4), with test administration strategies that replicate conditions prevalent during emergency conditions, such as (1) extreme time pressure; (2) ambiguous instructions, (3) contextual information that changes significantly over time, sometimes quickly; and (4) provision of relevant and irrelevant information at the same time. In other words, the testing format is configured to represent conditions and challenges that are frequently present in real-world emergency situations.

APPENDIX A
WATER UTILITY SURVEY QUESTIONNAIRES

The Stratus Consulting team conducted a two-part survey of water utilities, including representatives from large, medium, small, and predominantly rural systems. This appendix contains the two questionnaires: (1) a 46-question, yes/no answer questionnaire delivered and returned via e-mail; and (2) a 31-question telephone survey addressing a wide range of emergency planning topics, typically requiring approximately 1 hour to complete.

Disaster response, recovery and business continuity planning for drinking water utilities - AwwaRF Project 2929

WATER UTILITY SURVEY
(March 23, 2004)

Survey Purpose:

Thank you for participating in this confidential survey. It should take about 5 minutes to complete. It is part of an AwwaRF research project (2929) to characterize the state-of-the-practice for water utility disaster response, recovery and business continuity planning in the United States in order to develop a template for the water industry that shows best practices in this area. At the conclusion of the project, we will summarize our findings and identify significant opportunities where practice enhancements can be made to disaster planning in water utilities. The report will be made available through AwwaRF.

Definitions:

Response – Disaster response takes place at the incident stage. It consists of activities conducted to save lives and prevent harm to people and property during an emergency. The immediate response comprises; spreading the alert, establishing control, and mobilizing staff.

Recovery – Disaster recovery is defined as activities that restore vital life-support systems to minimum operating standards, and return infrastructure systems to a state of normalcy. The disaster recovery process begins during the response phase and as soon as the consequences of the incident become clear.

Business Continuity – Business continuity is defined as activities that ensure that the water utility is able to continue its business of providing water services to its customers until business functions assume a state of normalcy. Business continuity activities start during the response phase and end after the utility has resumed normal operations.

Utility information	
Date:	
Utility/state:	
Population served by utility:	
Name/position of respondent:	
Length of employment with utility:	
Phone number/email of respondent:	

No.	QUESTION	YES	NO
	Characterization of disaster/emergency scenarios		
1	Has your utility experienced any man-made/human-caused disaster during the past 10 years that could have resulted in a water system interruption? If yes, indicate type of disaster in the following list:		
2	Flood/storm disaster?		
3	Seismic disaster?		
4	Other? (please specify)		
5	Has your utility characterized potential emergency/disaster scenarios from a planning standpoint (e.g., man-made or natural, or based on level of severity)?		
	Vulnerability Assessment		
6	Has your utility performed a vulnerability assessment (VA) in response to the new EPA Requirements? If yes, when?		
7	Did The VA include an IT security assessment?		
8	Did the VA include a threat assessment?		
9	Has it been updated within the last two years?		
10	Does your organization's policies require periodic assessments?		
11	Do you have tools for real-time contaminant detection in your distribution system?		
12	Does your utility have any criteria for redundant facilities to handle emergencies or disasters?		
13	Do you have a SCADA system?		
14	Do you deny access to the general public to your critical facilities, such as water treatment plants, without escort?		
15	Do you give access to the general public to your watershed for recreation?		
	Emergency Response Plan (ERRP)		
16	Does your utility have an emergency response plan that includes a plan for system recovery (ERRP)?		
17	Are all of your response and recovery plans/protocols in one document?		
18	Does organizational policy require an update/revision to your ERRP after completion of the VA?		
19	Do you require that the ERRP be updated every year?		
20	Did your utility use any emergency planning and /or disaster management software tools in developing its ERRP or other emergency preparedness activity?		
21	Did you use any emergency management standard or guideline such as Standardized Emergency Management System (SEMS) and/or AWWA-		

2

Disaster response, recovery and business continuity planning for drinking water utilities - AwwaRF Project 2929

No.	QUESTION	YES	NO
21	Did you use any emergency management standard or guideline such as Standardized Emergency Management System (SEMS) and/or AWWA-M19 in the development of the ERRP?		
22	Does your utility have an Emergency Response Manager/Coordinator?		
23	Is your utility involved in any inter-organizational cooperative mechanisms between water agencies, such as mutual assistance?		
24	Is your utility involved in any cooperative mechanisms between water agencies at the state level?		
25	Is your utility involved in any cooperative mechanisms between water agencies at the regional level?		
26	Does your ERRP provide directions for personnel safety?		
27	Is the organizational structure of your ERRP consistent with the Incident Command System used by other responders (e.g., fire, police) in your area?		
28	Has your utility's ERRP ever been activated since it was developed?		
29	Is your utility involved in any cooperative mechanisms between water agencies at the local level?		
	Emergency Recovery Plan (ERRP)		
30	Was your ERRP developed around a worst-case man-made disaster scenario?		
31	Was your ERRP developed around a worst-case flood/storm disaster scenario?		
32	Was your ERRP developed around a worst-case seismic disaster scenario?		
33	Does your ERRP identify and characterize system components?		
34	Does it identify alternative water sources?		
35	Does your ERRP identify equipment that can be used to mitigate or lessen impact of an emergency situation?		
36	Does your emergency response plan include incident specific emergency action field procedures?		
37	Is the distance from the nearest police/emergency response agencies known?		
38	Have the known or likely response times of police and emergency response personnel been determined?		
	Business Continuity Planning		
39	Does your organization have provisions, procedures, plans under the category of business continuity?		
40	Has your business continuity plan ever been used?		
41	Does your organization have a schedule and budget for updating and on-going plan maintenance activities?		
	Training and Drills		

3

No.	QUESTION	YES	NO
42	Have the emergency personnel in your utility been given any training on the ERRP?		
43	Do you activate the ERRP for practice drills?		
44	Do you discuss emergency response strategies for practice scenarios among staff on a regular basis?		
45	Do you practice emergency response activation with outside agencies?		
	General		
46	How do you rate your emergency response preparedness on the following scale? Please circle only one: 1 – Excellent 2 – Good, might be enhanced 3 – Fair, needs a number of improvements 4 – Improvement needed, needs a number of enhancements 5 – Not adequate	1 2 3 4 5	

DISASTER RESPONSE, RECOVERY AND BUSINESS CONTINUITY PLANNING FOR DRINKING WATER UTILITIES
AWWARF PROJECT 2929

WATER UTILITY INTERVIEW GUIDE
(March 29, 2004)

Utility information

Date:

Utility/state:

Utility identification number:

Name/position of respondent:

Length of employment with utility:

Phone number/email of respondent:

Population served by utility:

Characterization of disaster/emergency scenarios

1. What kinds of emergencies/disasters has your utility had over the last 10 years?

2. How does your utility distinguish between an emergency and a disaster? Which criteria are used in each case? (e.g., number of people out of service or affected).

3. Please explain the emergency/disaster scenario characterization being used by your utility.

Vulnerability Assessment (Note to interviewer- Most utilities will be hesitant to answer too many specifics in this area, however since the focus of the project is on disaster response, recovery, and business continuity it is not necessary to have much detail here).

4. What are the perceived threats your utility faces in order of importance? When were these last determined? Does the assessment include cyber threats?

1

5. How does your utility detect and identify threats?

Emergency Response and Recovery Planning

6. Please state the main elements of your ERRP (Note - be sure to make clear the distinction between a response plan and a recovery plan and determine which one or both the utility has)?
7. What are your criteria for establishing system redundancy for emergencies?
8. Please describe the emergency scenario/s that your ERRP was based on?

9. What are the outside parties that your utility's ERRP has been distributed to?

10. Please state any emergency planning and /or disaster management software tools used by your utility in developing its ERRP or other emergency preparedness activity?

11. How large is your ERRP document?

12. State and describe any emergency management standard or guideline such as Standardized Emergency Management System (SEMS) and/or AWWA-M19 used by your utility in the development of the ERRP?

13. What equipment does your ERRP identify that can be used to mitigate or lessen the impact of an emergency/disaster?

14. Does your plan include Facility Level Procedures for your utility?

15. What particular aspects of your utility's inter-organizational/inter-regional cooperative (e.g., mutual assistance) for emergency response and recovery seem especially useful?

16. Was any aspect of the agreement particularly ineffective or difficult to implement?

17. What are the criteria used for the activation of your ERRP?

18. In your experience, what is the single most important aspect of an effective emergency response and recovery plan?

19. Is there a utility or organization that you think of as "the best" or a model for emergency response, disaster management, or business continuity?

20. Does your ERRP have any emergency planning/disaster management communications plans or elements?

2

21. How does your plan characterize key internal and external stakeholder audiences and what are the key audience elements (e.g., businesses, residents, schools, media, politicians, Board Members, employees, vendors, etc.). How many audiences are included and what is the method and timing of contact when there is an emergency?

22. How useable is your plan in terms of how large, and do you feel confident that it can easily be implemented?

Business Continuity Planning

23. How does your utility define business continuity planning?

24. What are the elements of your business continuity plan? (Ask for Table of Contents).

25. What is your frequency for updates and maintenance activities? How does this compare to other similarly-sized utilities (in your region)?

Training Exercises and Drills

26. How many training exercises and drills are planned for on an annual basis? Can you describe the format of the training sessions?

27. Can you describe the format for the drills and table top exercises?

28. Can you please send me copies of the training materials and any pertinent information? (Give address, fax number, email number as appropriate – list what is being sent)

General

29. What aspect of your emergency/disaster response plan would you like to change?

30. Do you think this survey is missing any important questions?

31. Is there any else you would like to say?

Follow-up actions

Refer to questions 23 and 27.

3

APPENDIX B
KIT OF TOOLS:
ACTION AREAS AND SELF-ASSESSMENT QUESTIONS

Action Areas and Self-Assessment Questions

The following checklist identifies each of the 12 action areas described in chapter 3 of the report and the self-assessment questions that correspond to each area. The CD-ROM associated with this report contains resources and best practice information indexed to the self-assessment questions.

Water systems can use the checklist as a *self-assessment tool* to review and critique their emergency response plans. If, in reviewing a particular area, water system officials suspect that their existing plan may be lacking or insufficient with respect to a particular question, they can refer to the CD-ROM and "click through" to obtain relevant, detailed guidance.

Action Area 01—Recognize Status Changes and Activate Plan

A. How is a threat deemed credible?
B. Who should be contacted in case of an emergency?
C. How is an emergency reported and what information should it include?
D. How are different types of threats classified?
E. What factors determine or control activation of the emergency response and recovery plan?

Action Area 02—Mobilize Staff

A. Has the utility established an emergency operations center or otherwise designated space for the response team to meet during a crisis?
B. Has the utility written a crisis communications plan? Does the plan include pre-scripted response information dealing with likely scenarios, answers to frequently asked questions, and summary background information on possible crisis threats?
C. Is there a pre-assigned crisis team or similar group? Do team members have defined roles and responsibilities?
D. Does the plan include an updated list of contact information for team members, top management, and potentially relevant staff and contractors?
E. Do staff and contractor listings include multiple, diverse contact modes (e.g., phone, cellular phone, e-mail, pagers, radio call numbers, home addresses)?
F. Do all applicable staff members have either full copies or applicable sub-sections/summaries of the emergency response plan?
G. Have all employees been recently briefed on the emergency response plans, emergency management authorities and responsibilities, and key emergency procedures?
H. Has the crisis team drilled or conducted exercises in which the plan has been activated and employed?
I. Has the organization designated a primary and backup spokesperson?

Action Area 03—Activate External Communications

A. Is the plan based on thorough consideration of external stakeholders? Does the analysis of key stakeholders clearly differentiate how different emergency scenarios will affect different stakeholders?
B. Does the plan include provision for external aid or mutual assistance, if required? Does the utility have an active mutual assistance arrangement with neighboring utilities or other applicable entities?
C. Does the utility maintain a regular dialogue with relevant emergency response partners and stakeholders?
D. Does the community water system involve emergency response partners (e.g., fire, law enforcement, emergency medical) in training exercises?
E. Does the plan include crisis communications lists?
F. Does the plan include crisis communication templates that help to determine who to contact first, in what order to contact key stakeholders, frequency of information updates, and pre-scripted messages keyed to particularly likely scenarios?
G. Does the plan identify a media relations officer (spokesperson), a backup media relations officer, and an emergency communications team?
H. Do all staff members know or have access to the name and contact information for the designated media relations official?
I. Have designated spokespersons received training in how to deal with the media?
J. Does the plan include a list of customers with special needs? Have steps been taken to provide ample/adequate notification and information to customers with special communication needs (e.g., non-English speaking consumers, individuals with physical disabilities, individuals with limited reading skills)?
K. Does the plan address multiple methods to deliver public messages and notifications (e.g., radio, phone banks, TV "scrollers," door-to-door canvassing, broadcast faxes, network and cable television, special Web sites)?

Action Area 04—Analyze the Type and Severity of an Event

A. Does the plan include a list of critical system components?
B. Does the plan include a "damage checklist," including critical system components and likely damage modalities?
C. Does the plan include a decision tree, process schematic, worksheet, or similar tool to guide the threat evaluation process?
D. Does the plan (or an annex) contain a working list of potential contaminants, associated health and environmental impacts, and potential response/mitigation actions?
E. Does the plan include a decision tree, process schematic, or similar tool to guide analyses for potentially contaminated water samples?
F. Does the plan include a decision tree, process schematic, worksheet, or similar tool to guide site characterization activities, particularly to identify and characterize potential hazards to utility staff and other emergency responders?
G. Does the plan contain guidelines for maintaining crime scene integrity?
H. Does the plan contain—or direct responders to—water sampling guidelines and information on water contaminants, including non-traditional drinking water

contaminants? Does the plan contain a sample collection, identification, and chain-of-custody form?

I. Does the plan direct emergency responders to experts, modeling tools, and other resources to help assess the flow of contaminants in the system?

Action Area 05—Develop Specific Response and Recovery Actions

A. Does the plan clearly articulate the water system's goals for emergency response? Are the goals listed in order of priority?

B. Does the emergency response plan articulate step-by-step processes for evaluating and then addressing the spectrum of threat scenarios that the utility is likely to face? Do response protocols in the emergency response plan differentiate between "confirmed" and "non-confirmed" incidents?

C. Is the response planning template based on the community water system's most recent vulnerability assessment or other carefully structured, fact-based inputs?

D. Does the plan include a "priority service form" or similar mechanism for ensuring that critically dependent customers receive an uninterrupted supply of water?

E. Does the plan (or annex) contain a set of response procedures for likely scenarios and events? Do response strategies focus on immediate containment or isolation of the event?

F. Do response protocols include provision to obtain and utilize alternate water supplies? If "yes," is it clear that alternative sources have been identified? Are mutual aid or other agreements already in place to authorize and initiate provision and delivery of alternative water supplies?

Action Area 06—Assess and Address Health and Safety Issues

A. Do response protocols clearly differentiate between known, unknown, and suspected contaminants?

B. Does the plan contain a public health response action worksheet that addresses (1) identification of the contaminant, (2) contaminant properties, (3) containment options, and (4) critical equipment within the contaminated area?

C. Does the plan (or annexes) contain summary information on properties applicable to biological (e.g., minimal infective dose, secondary transmission, susceptibility to disinfection, survivability in water) and chemical (toxicity values, aerosolization, hydrolysis, solubility, reactivity) contaminants?

D. Does the plan outline site control measures?

E. Does the plan include a description of personnel protective equipment and its location?

F. Does the plan include step-by-step decontamination procedures for likely contaminants?

G. Does the plan (or annex) contain sample notification and treatment forms, such as "water shutoff," "boil water before using," "do not drink water," and "do not use" notifications? If "yes," do these notifications (1) tell the recipient what to do, (2) describe what happened and what is being done, and (3) provide a contact source for additional information?

H. Does the plan contain a table of important translated phrases—such as "do not drink," "boil water before using," and "do not use water"—in locally applicable languages?

Action Area 07—Preserve Stored Water and Identify Alternative Water Sources

A. Does the plan (or annex) include a list of options for short term alternate supplies of drinking water, characterizing both type (e.g., bottled, bulk, untreated) and source (e.g., local government agencies, local retailers, military assets, neighboring water utilities)?
B. Is it clear that alternate supplies are substantiated through current mutual aid agreements, contractual arrangements, or other mechanisms?
C. Does the plan include information on state and federal agencies (e.g., U.S. Army Corps of Engineers, FEMA) that may be called on if local sources of water prove unavailable or insufficient?
D. Does the plan contain fact sheets and/or instructions that consumers can use to identify household sources (e.g., pipes, hot water tanks) of emergency water supply?
E. Does the plan include fact sheets or other public guidance for water purification?

Action Area 08—Perform Emergency Repairs Based on Priority of Demand

A. Does the plan (or annex) include a list of potential repair and reconstruction support needs, including equipment, suppliers, contractors, lifeline support, and cooperating agencies (e.g., electric utilities, wastewater utilities)?
B. Does the plan include guidance on how to dispose or otherwise manage different types of debris?
C. Does the plan (or annex) include pre-formatted work-order logs?

Action Area 09—Return System to Normal Levels

A. Does the plan include or direct users to a recovery period checklist or other guidance on how to conduct a detailed post-incident system characterization and residual risk assessment? Does this post-incident characterization process include guidance on how to coordinate with appropriate civil authorities and insurance underwriters?
B. Does the plan (or annex) include a decision tree or similar tool that addresses key restoration issues, such as long-term alternate water supplies (which may be different than short-term water supply), analysis of candidate cleanup/remediation approaches, preparation of remedial designs, and implementation of cleanup/remediation activities?
C. Does the plan contain an incident command structure for remediation activities?
D. Does the plan direct users to USEPA guidance and resources concerning remediation and recovery, including the agency's guidance for process documentation?
E. Does the plan include templates for post-event public communication?

Action Area 10—Evaluate Response and Plan

A. Does the plan include forms and instructions for documenting staff activity and expenses incurred in response to a declared disaster?
B. Does the plan include sample forms and/or instructions for developing an incident report?
C. Does the plan include guidance on the preparation of a post-event evaluation report? Does guidance address issues, background, recommendations, lessons learned, and conclusions?

Action Area 11—Revise Plan

A. Does the plan (or annex) contain a distribution list, indicating when specific individuals were provided with copies of the plan?

B. Does the plan (or annex) include a record of changes, including numbered changes, description of change, section(s) or pages affected (as applicable), date of change, and the name of the individual who authorized the change?

Action Area 12—Conduct Ongoing Training and Security Acculturation

A. Does the plan include a regular training component? Does this section (or annex) articulate who will be (was) trained, who will do (did) the training, what training approaches/activities will be (were) utilized, and when and where training sessions will take (took) place?

B. Do (did) training activities involve applicable stakeholders and emergency response partners?

C. Does the plan or associated program distinguish among initial training, detailed assignment training, and ongoing refresher training?

D. Does training include necessary certifications (e.g., handling hazardous chemicals)?

E. Does the program include a method for testing the plan itself?

F. Do training exercises mesh with key vulnerabilities and/or threat scenarios as articulated in the water system's emergency plan and/or vulnerability assessment?

G. Is there evidence that the plan is becoming integrated with overall utility operations? Does senior management promote responsibilities outlined in the plan? Do employee evaluations include emergency management responsibilities? Have emergency planning concepts been incorporated into accounting, personnel, and financial operations? Are emergency response materials distributed through corporate newsletters, employee mailings, and other vehicles?

REFERENCES

Alexander, D. 2002. *Principles of Emergency Planning and Management*. New York: Oxford University Press.

Anderberg, K., C. Bersh, S. Kelly, and R. Peckham. 2002. Disaster Recovery System Ensures 100% Uptime. *Communications News*, 39(1):8.

Andrews, R. 2002. A Utility's Security Game Plan Must Include Crisis Management. *Electric Light & Power*, 80(11):10.

AWWA (American Water Works Association). 2001a. Emergency Planning for Water Utilities—Manual of Water Supply Practices M19. Denver, Colo.: AWWA.

AWWA (American Water Works Association). 2001b. Security Analysis & Response for Water Utilities. Denver, Colo.: AWWA.

AWWA (American Water Works Association). 2002. Water System Security: A Field Guide. Denver, Colo.: AWWA.

Babiuch, W., and B. Farhar. 1994. *Stakeholder Analysis Methodologies Resource Book*. NREL/TP-461-5857. National Renewable Energy Laboratory.

Baddeley, A. 1992. Working Memory. *Science*, 255(5044):556-559.

Chenn, A., and M. Hicks. 2001. Testing Your Disaster Recovery Plan. *eWeek*, 49, Oct. 1.

Community Banker. 2003. FFIEC Guidance for Business Continuity Planning.

CPM (Contingency Planning and Management). April 2003. Essential Ingredients of a BC Plan. *Contingency Planning & Management,* 8(3):8 Available: <http://www.contingencyplanning.com/archives/2003/apr/8.aspx>.

DHS. 2004. National Incident Management System. Available: <http://www.fema.gov/nims/>.

Dwyer, P.D., A.H. Friedberg, and K.S. McKenzie. 1993. It Can Happen Here: The Importance of Continuity Planning. *National Public Accountant*, 38(6):June.

FEMA. 1993. *Emergency Management Guide for Business & Industry*. Available: <http://www.fema.gov/pdf/library/bizindst.pdf>.

FEMA. 2002. *Toolkit for Managing The Emergency Consequences Of Terrorist Incidents*. Available: <http://www.fema.gov/preparedness/toolkit.shtm>.

FEMA. 1996. *Guide for All-Hazard Emergency Operations Planning*. Available: <http://www.fema.gov/rrr/gaheop.shtm>.

FEMA. 2001. *Guide for All-Hazard Emergency Operations Planning*. Available: <http://www.fema.gov/pdf/rrr/allhzpln.pdf>.

FEMA. 2002. *Managing the Emergency Consequences of Terrorist Incidents – Interim Guidelines*. Available: <http://www.fema.gov/pdf/onp/managingemerconseq.pdf>.

FEMA. 2003. *Federal Response Plan*. Available: <http://www.fema.gov/pdf/rrr/frp/frp2003.pdf>.

Grabowski, M., and K. Roberts. 1997. Risk Mitigation in Large-Scale Systems: Lessons From High Reliability Organizations. *California Management Review*, 39(4):Summer.

Grigg, N. 2002. *Surviving Disasters in Water Utilities: Learning From Experience*. Denver, Colo.: AwwaRF and AWWA.

Hopkins, P. 2004. Business Continuity Planning and Water Utilities. Presented at AwwaRF Project 2929 Project Workshop, San Francisco, Calif.

Hynes, T., and P. Prasad. 1997. Patterns of Mock Bureaucracy in Mining Disasters: An Analysis of the Westray Coal Mining Explosion. *Journal of Management Studies*, 12:76-90.

IDEQ (Idaho Department of Environmental Quality). 2001. *Public Water System Emergency Response Plan Guide*. Available: <http://www.deq.state.id.us/water/dw/EmergencyResponseForm.doc>.

Kirvan, P. 2003. Business Continuity Plan Exercising: What You Need to Know. *Contingency Planning & Management*, 8:7:5.

Lavie, N. 1995. Perceptual Load as a Necessary Condition for Selective Attention. *Experimental Psychology: Perception and Performance*, 21(3):451-468.

Little. R. 2004. A Holistic Strategy for Urban Security. Unpublished manuscript. Washington, D.C.: National Research Council.

Lustberg, A., and B.R. Silvenberg. 1995. Sending the Right Message When Crisis Strikes: Nine Steps to Preventing a Communication Nightmare. *Association Management*, 47:July.

Markwood, I., and W.F. Nabak. 1993. Careful Emergency Planning Will Pay Dividends. *Water Engineering & Management*, 140(5):May.

Metropolitan Washington COG (Council of Governments). 2002. *The Regional Emergency Coordination Plan*. Washington, D.C. Available: <http://www.mwcog.org>.

Michael Baker Corporation. 2004. *Security Practices Primer for Water Utilities*. Denver, Colo.: AwwaRF.

Mitroff, I.I., and M.C. Alpaslan. 2003. Preparing for Evil. *Harvard Business Review*, April:109–115.

Moore, P. 2000. The Umbrella of Business Continuity. *Canadian Insurance*, 10(9):21-24.

Mutual Assistance Agreement. 1982. Various utilities in the Western United States.

NeRWA (Northeast Rural Water Association). 2002a. *Rural & Small Water and Wastewater System Emergency Response Plan Template*. Available: <http://www.neruralwater.org/downloads/NeRWA_EmergencyResponseTemplate_Nov2002B.pdf>.

NeRWA (Northeast Rural Water Association). 2002b. *Water System Emergency Response Plan*. Available: <http://www.neruralwater.org/downloads/NeRWA_ERP.pdf>.

NETC (National Environmental Services Center for Small Communities). 2002. *Protecting Your Community's Assets: A Guide for Small Wastewater Systems*.

Newell A., and H.A. Simon. 1972. *Human Problem Solving*. Englewood Cliffs, N.J.: Prentice Hall.

Office of Emergency Services (California). 1999. *Emergency Planning Guidance–Public and Private Water Utilities*. Available: <http://www.oes.ca.gov/oeshomep.nsf/all/WaterUtilities/$file/H2o_.pdf>.

Office of Emergency Services (California). 2001. *California Terrorism Response Plan–An Annex to the State Emergency Plan*. Available: <www.oes.ca.gov/oeshomep.nsf/all/TerrorPlan/$file/TerrorPlan.pdf>.

Pearson, C.M., and J.A. Clair. 1998. Reframing Crisis Management. *Academy of Management Review*, 23(1):January.

Public Health Security and Bioterrorism Preparedness and Response Act of 2002 (Public Law 107–188).

Schulman, P., E. Roe, M. van Eeten, and M. de Bruijne. 2004. High Reliability and the Management of Critical Infrastructures. *Journal of Contingencies and Crisis Management*, 12(1):14–28.

SDWA (Safe Drinking Water Act). Amendments: section 1433 (42 USC 300i-2).

Smith, D. 2003. Learning From BCI's God Practice Guidelines: Business Continuity Advice From Someone Who Knows. *Contingency Planning and Management Magazine*, 8:4:2. Available: <http://www.contingencyplanning.com/PastIssues/apr2003/2.asp>.

Swichtenberg, B. 2002. Staying Alert. *Water Engineering & Management*, 149(6):June.

Tufte, E.R. 2001. *The Visual Display of Quantitative Information*. 2nd ed. Cheshire, Conn.: Graphics Press.

U.S. EPA. 2000. *Publication Notification Handbook.* Available: <www.epa.gov/safewater/pws/pn/handbook.pdf>.

U.S. EPA. 2002. *Guidance for Water Utility Response, Recovery & Remediation Actions for Man-Made and/or Technological Emergencies*. Available: <http://www.epa.gov/safewater/watersecurity/pubs/er-guidance.pdf>.

U.S. EPA. 2003a. *Instructions to Assist Community Water Systems in Complying with the Public Health Security and Bioterrorism Preparedness and Response Act of 2002*. Available: <http://www.epa.gov/safewater/watersecurity/pubs/util-inst.pdf>.

U.S. EPA. 2003b. *Large Water System Emergency Response Plan Outline: Guidance to Assist Community Water Systems in Complying with the Public Health Security and Bioterrorism Preparedness and Response Act of 2002*. Available: <http://www.epa.gov/safewater/watersecurity/pubs/erp-long-outline.pdf>.

U.S. EPA. 2003c. *Response Protocol Toolbox: Planning for and Responding to Drinking Water Contamination Threats and Incidents – Overview and Application*. Interim Final – December 2003. Available: <http://www.epa.gov/safewater/security/pdfs/guide_response_overview.pdf>.

U.S. EPA. 2003d. *Response Protocol Toolbox: Planning for and Responding to Drinking Water Contamination Threats and Incidents – Module 1: Water Utilities Planning Guide.* Interim Final – December 2003. Available: <http://www.epa.gov/safewater/security/pdfs/guide_response_module1.pdf

U.S. EPA. 2003e. *Response Protocol Toolbox: Planning for and Responding to Drinking Water Contamination Threats and Incidents – Module 2: Contamination Threat Management Guide*. Interim Final – December 2003. Available: <http://www.epa.gov/safewater/security/pdfs/guide_response_module2.pdf>.

U.S. EPA. 2003f. *Response Protocol Toolbox: Planning for and Responding to Drinking Water Contamination Threats and Incidents – Module 3: Site Characterization and Sampling Guide*. Interim Final – December 2003. Available: <http://www.epa.gov/safewater/security/pdfs/guide_response_module3.pdf>.

U.S. EPA. 2003g. *Response Protocol Toolbox: Planning for and Responding to Drinking Water Contamination Threats and Incidents – Module 4: Analytical Guide*. Interim Final – December 2003. Available: <http://www.epa.gov/safewater/security/pdfs/guide_response_module4.pdf>.

U.S. EPA. 2003h. *Response Protocol Toolbox: Planning for and Responding to Drinking Water Contamination Threats and Incidents – Module 5: Public Health Response Guide.* Interim Final – December 2003. Available: <http://www.epa.gov/safewater/security/pdfs/guide_response_module5.pdf>.

U.S. EPA. 2003i. *Response Protocol Toolbox: Planning for and Responding to Drinking Water Contamination Threats and Incidents – Module 6: Remediation and Recovery Guide.* Interim Final – December 2003. Available: <http://www.epa.gov/safewater/security/pdfs/guide_response_module6.pdf>.

U.S. EPA. 2004. *Emergency Response Plan Guidance for Small and Medium Community Water Systems to Comply with the Public Health Security and Bioterrorism Preparedness and Response Act of 2002*. Available: <http://www.epa.gov/safewater/watersecurity/pubs/small_medium_ERP_guidance040704.pdf>.

Verton, D. 2003. Blogs Play a Role in Homeland Security: Make it Easy for Critical-Infrastructure Agencies to Share Info From Bottom Up. *Computerworld*, 37(19):12(1).

Wade, B. 2002. Locking Down on System Security. *American City & County*, 117(1):January.

Water Week. 2004. 13(41). October 6.

West Virginia University, National Environmental Services Center for Small Communities. 2002. *Protecting Your Community's Assets: A Guide for Small Wastewater Systems*.

ABBREVIATIONS

AWWA	American Water Works Association
AwwaRF	Awwa Research Foundation
BCP	business continuity planning
COG	Council of Governments
CPM	Contingency Planning and Management
DHS	Department of Homeland Security
EPCRA	Emergency Planning & Community Right to Know Act
ERRP	Emergency Response and Recovery Plan
FEMA	Federal Emergency Management Agency
HROs	high-reliability organizations
IT	information technology
LEPCs	Local Emergency Planning Committees
NIMS	National Incident Management System
PAC	project advisory committee
PDA	Personal Digital Assistant
SCADA	supervisory control and data acquisition
SDWA	Safe Drinking Water Act
SEMS	Standardized Emergency Management System
SOPs	Standard Operating Procedures
USEPA	U.S. Environmental Protection Agency

IWA Publishing's authorised EU representative for General Product Safety Regulations is Diane D'Arras, 15 rue Duret, 75116 Paris, France, e-mail: safety@iwap.co.uk.

Printed and bound by CPI Group (UK) Ltd, Croydon, CR0 4YY
17/09/2026
02218810-0008